T0264504

CRC Press
Taylor & Francis Group
6000 Broken Sound Parkway NW, Suite 300
Boca Raton, FL 33487-2742

First issued in hardback 2017

ISBN 13: 978-1-138-41234-7 (hbk)
ISBN 13: 978-1-57820-026-9 (pbk)

Visit the Taylor & Francis Web site at
http://www.taylorandfrancis.com

and the CRC Press Web site at
http://www.crcpress.com

MAXIMIZING

CALL CENTER PERFORMANC

136 INNOVATI

IDEAS FOR INCREASII

PRODUCTIVITY A

CUSTOMER SATISFACTI

CRC Press
Taylor & Francis Group
Boca Raton London New York

CRC Press is an imprint of the
Taylor & Francis Group, an **informa** business

BY MADELINE BODIN

CONTENTS

INTRODUCTION

HOW TO USE THIS BOOK

Call centers are versatile tools. They can provide sales, service, technical support or just information for your customers and prospects. They can make your customers feel your company is caring, efficient, informed, cutting-edge, fun — or give them a much different feeling.

How you use your call center — and how your customers feel about it — is up to you, and how you and your company can leverage call center technology to serve your needs.

But if you are isolated in your own call center, how can you know what options are available? This book can help. This book can show you how companies like yours solved problems like yours. It can show you how companies from other industries put together technology solutions for applications that your company performs too.

Learn from them.

This book is organized by application: billing, collections, help desk, sales, customer service and technical support. But the stories included in this book are true stories and companies must perform many applications with their call centers. So the companies are categorized under the application that is best illustrated in the story. That doesn't mean there aren't great customer service ideas under sales. Or great ways to distribute information under technical support. Keep an open mind and browse a little.

If you are looking for a solution for a particular call center problem, don't limit yourself just to the chapter on your specific application. There is no firm line that distinguishes sales from order handling. You might find your solution in either chapter.

Under each application, the stories are categorized first by industry, then by technology. Among the industries profiled in this book are:

banking	cable
catalogs	collections agencies
financial services	government
high tech	non-profit/fundraising
packaged goods	publishing
retail	travel
utilities	

It may be possible to zero in on how another company in your industry solved its problem with (for example) order handling. You may find several solutions, each using a different technology. Or you may find that particular industries keep turning to the same technologies again and again. Each story will illustrate a different method of using that technology, however.

Once again, you may not find an exact fit with your industry, application or problem. Don't despair. Browse. Read stories that are similar to your own. Read stories that catch your eye. The solution may be here in this book, but I may have categorized it differently than you would.

Some of the best answers to your call center's problems are not neatly catagorized anyway. A flash of insight may show you how a technology that solved one company's information distribution problem is the perfect solution for your sales dilemma.

Finally, I have showcased several cutting-edge call center technologies by giving them their own chapters. The applications for these technologies — often few in number — offer a look into a crystal ball. They show you what your call center can be like in the future — or tomorrow.

The technologies honored with special chapters are:

call blending	computer telephone integration
the Internet	kiosks
virtual call centers	

The chapter on monitoring and agent evaluation could probably be put in either category — application or technology. In this case, the two are one and the same.

The tables at the end of the book provide a handy map to the solutions that best fit your situation. The first table lets you find how others in your industry used call center technology to automate a particular function, such as sales or a help desk. The second table lets you see how various technologies (such as Automatic Call Distributors or Interactive Voice Response systems) have been used in and across industries.

Last but not least, there is an index listing solutions by technology vendor. Want to see how other companies are using the technology you already have? Check here. Want to see how the vendor you are thinking of buying from has performed in the past? Look it up.

I hope this book gives you ideas that you can put to work in your call center immediately. It is a treasure trove of what some of the best call centers out there have done — with hints at what they plan to do in the future.

This book has 136 great call center applications. How many does your call center need? If it makes your customers happy and your company successful, just one.

MADELINE BODIN

1

BILLING
AND CREDIT

CABLE
CALL CENTER MANAGEMENT

Sending overflow calls from a center in say, Chicago to another center in Salt Lake City sounds very sophisticated and cutting edge. It is. But the same amount of sophistication and cutting edge technology — or more — is required to send overflow calls from one end of a metropolitan area to the other.

That's what a cable company is doing in their two call centers — one in a major city, the other in a nearby suburb.

The two centers perform the same functions, but serve different geographical areas, says the cable company's former VP of telecommunications and customers service, who is working with the cable company as a consultant on this project.

The suburban center has 74 agents and the city center has 40. The suburban center not only has more agents, but they also receive more calls than the city center. A lot more.

Recent changes in rates and channel realignments means the percentage of customers calling in to the suburban center each month jumped from 15% to nearly 100%. The ability to flow calls from the overwhelmed suburban center to the city center has helped the cable company maintain their customer service levels.

"It's a very sophisticated environment," notes the consultant. Automatic number identification (ANI) and dialed number identification service (DNIS) are provided by US West and are delivered to the center through the primary rate interface (PRI) of ISDN (integrated services digital network).

The company is still working on a computer-telephone integration (CTI) application where the ANI and DNIS information will prompt screen-pops at the agents' desks.

"Currently our billing system doesn't allow true CTI, so we don't have screen pops," says the consultant. "But all the other elements are there to allow dynamic call handling."

One key part of the two-center system is the Call Center Solutions call center management software from Chadbourn Marcath.

The Call Center Solutions system communicates over a fiber optic link between the two centers.

"It was important that we could integrate both centers and get a data stream from both centers," says the consultant. The Call Center Solutions packages monitors data pumped from the two Rolm ACDs that serve the cable company call centers.

The consultant says the Call Center Solutions system is a key part of the cable company's ability to dynamically overflow calls between centers. "We worked with US West and Rolm to regulate our trunks and our call flow. The Call Center Solutions system helps us regulate and monitor the information."

As the cable company works on eliminating the bottleneck created by their billing system, the Call Center Solutions package is letting them handle calls efficiently by distributing them between the two locations.

"We are patiently waiting for true CTI," says the consultant, "but Call Center Solutions is filling the gap with call-flow handling."

CREDIT
PREDICTIVE DIALING

A check verification and recovery agency wanted to reduce the time spent dialing calls manually, and reduce the paper work of making notes on each call attempt and automate its manual system.

It used a predictive dialing system from EIS.

Using the system, the number of connected callers was doubled, to 50 per hour per agent. The number of calls attempted per hour per agent also more than doubled, from 40 to 100. The number of man-hours of collection personnel has been reduced by more than 50%.

The dialer is linked to the company's host computer. Not only is a computer record brought to the agent's screen with each call, but all dial attempts are recorded in the host computer files.

Inspired by its collections success with the EIS system, the company put the dialer to work producing sales leads. That program has also been very successful.

PREDICTIVE DIALING

One auto loan company's customers need special attention. This finance company specializes in high risk auto loans. It works with about 1,400 auto dealers to help customers, who might not otherwise be able to get a loan, buy a car.

The company's philosophy is that these customers may just need a little extra prodding to pay. Collections is a vital part of their operation.

A predictive dialing system from EIS helps the company give the special attention their customers need. The company calls accounts that are from three days to 59 days past due.

The company has been using the EIS system for about nine months. Previously collectors dialed accounts manually. In that time, the collection center manager says, "Outgoing call volume has increased, driving more money into the organization."

He reports that most of the collectors enjoy the faster pace of working with the predictive dialer. Few miss the wasted time spent listening to the telephone ring when there is no answer.

The loan company's collectors work from 8 AM to 8 PM from their call center in the Midwest. The EIS system's time zone technology makes it easier to call accounts from Maine to California at the right time of day.

The collection center holds nine licenses and has nine people on the phones non-stop during calling hours. There are three shifts and collectors usually work five or six hours a day.

The collection center manager credits excellent training by EIS for getting the center up to speed so quickly with the system. Other benefits include improved reporting and the ability to track collectors' work by looking at a computer screen of current information.

HEALTH CARE
ACD

A major health insurance company in Missouri has been using an Aspect CallCenter ACD for three years for a wide range of applications.

They use it to answer calls about membership and calls from health care providers. They use it in their billing department. One behavioral health subsidiary actually provides services by phone.

The big company has lots of offices in the region, but it once had to limit the use of its ACD to within a few thousand feet of the switch itself, located in their headquarters. Using remote shelves at the different sites was a possibility, but with the personnel using the switch thinly scattered among many locations, it just wasn't cost effective.

Recently though, the company began testing Aspect's WinSet for Windows, a PC application that delivers agent capabilities to any location that can be linked to the ACD through two standard telephone lines, an ISDN line or a wide area network. The new product has allowed the health care company expand the reach of its ACD beyond the walls of its headquarters.

These days, the company offers much more than health insurance. One if its many subsidiaries is a primary health care provider. WinSet let the company expand ACD capabilities to the patient service reps (receptionists) at four of the company's 30 health centers. (A health center functions like a doctor's or group practice's office.)

A call to one of the linked health centers is received by the ACD at headquarters and is routed to the appropriate center through dialed number identification service (DNIS). If that center is swamped with calls, or can't take the call for some other reason, the ACD routes the call to another health center, or it can be handled by one of the agents at the company's headquarters.

The health centers are located up to a 30-minute drive from headquarters and are each staffed with two or three patient service reps. Because health care is prone to emergency situations, it is not uncommon for the reps to become overloaded. The ACD link provides the reps with an orderly way to handle calls, plus a backup in case of emergency.

For management, the link provides a rich source of information. "It may be obvious to say this," says the company's telecom manager, stating what many managers fail to realize, "but if you try to manage without management information, you might as well be blindfolded."

She gives rep scheduling as an example. Ask a health center how many calls they get, and they may say they get "a lot" of calls on Mondays. "Is that three people or eight people?" asks the telecom manager. And are Mondays really busier, or do they just feel busier? ACD information answers these questions and lets the company use their resources effectively.

While the company is still just testing the WinSet system, the telecom manager already sees areas for expansion. Her first desire would be to get all the company's health care facilities linked so they can support each other during busy periods. She also sees the possibility of using home agents to help expand call center operations, especially in the marketing department, without taking up more expensive real estate.

Just as this insurance company has expanded far beyond its traditional role in the health care industry, WinSet has expanded the CallCenter ACD's role far beyond the traditional call center.

THE HEALTHCARE CALL CENTER

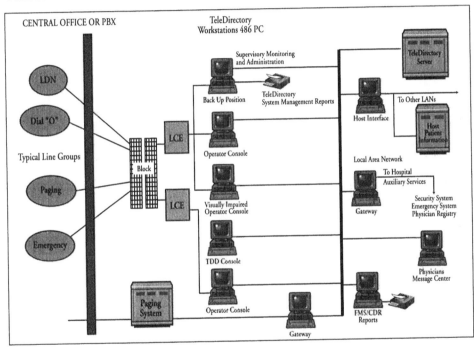

The hospital "call center" can use much the same technology as other call centers. They just use it in a very different way. These centers tend to have fewer agents, and those agents are often highly trained. Sometimes they are Registered Nurses. These agents must juggle a host of functions, including serving as a paging and message center for physicians. Some of these physician pages will be emergencies. Similarly, some of the calls coming into the "center" will literally be a matter of life and death, while others are routine requests to be routed to a patient's room. In these centers, good information — about physician availability, patient history and billing — is vital.

2

MEDIA BLENDING

BANKING
INBOUND/OUTBOUND

A bank in the southwest wanted to create universal agents who can be assigned to inbound or outbound calls as needed.

This bank used IMA's Telemar software, IBM's CallPath/400 and a predictive dialer from CAS.

When minimum service levels for the inbound queue are exceeded, the TSRs get a break message telling them to take an inbound after they finish their current call. Otherwise, the TSRs may choose to make another outbound call or take an inbound call after checking on the length of the inbound queue. The system uses ANI to display customer account information and the appropriate script on the TSR's screen.

COLLECTIONS
ACD/IVR

A collections agency wanted to blend inbound and outbound calls. To increase the number of debtors the company contacts daily and assure inbound calls are quickly routed to the assigned collector.

It used Rockwell's Spectrum ACD, Tandem's NonStop Himalaya servers and InterVoice's Voice Response Unit in its 60-collector shop.

Using the system, daily call placement volume increased over 250%, "right party" contacts increased 25%. Outbound call placement volume is four calls per second.

Collectors receive inbound calls and participate in outbound campaigns from the same station, even at the same time. The system captures incoming phone numbers and records them on the associated account.

INBOUND/OUTBOUND

A New England electric company wanted to handle increasing workload by switching to telephone-based collections and deal with seasonal peaks in call volume.

It used a predictive dialer from Davox and an ACD from Aspect.

Using the system, inbound call waiting times dropped 20%. Contacts per month increased from 2,000 with 18 agents to 4,500 with four agents. The company saves over $350,000 a year in salaries and benefits. The average amount collected per contact increased 80%.

The Davox dialing system monitors the number of inbound calls and stops making outbound calls when the wait-time passes a threshold. The system also routes overflow collections calls to customer service representatives.

The combination of a state law that prohibits cutting off people's electricity for non-payment between November 1 and April 15, and the fact that the company is in a college town, with students opening new accounts in September and closing them in May, means that spring and fall are very busy for the utility. The new system lets it deal with these busy seasons more efficiently.

ISDN

A collections agency wanted to improve performance by taking advantage of ISDN features.

Using a computer-telephone integration system, the agency does database lookups based on ANI, inbound call routing based on DNIS and inbound/outbound call blending.

When the agent is on an outbound call and an inbound call comes in, a window appears on the agent's screen with the caller's name and the amount of money they owe. If the incoming caller's debt is much higher than the debt associated with the outgoing call, the agent may wrap up the outgoing call quickly to take the inbound call.

HEALTH
IVR/FAX

A large southern HMO wanted to reduce call volume and handle the increased work load created when it was assigned to take over one-third of the state's former the Medicaid patients. The assignment increased their membership from 25,000 to 250,000.

It used a Micro-ITC System from Wygant Scientific which includes voicemail, IVR, fax-on-demand and voice recognition.

The Micro ITC System's IVR application processes 5,000 calls per month, siphoning off basic eligibility questions from the ACD queue. After entering information including their provider number and member ID, the system retrieves information from the company's mainframe. Callers can get fax copy of their eligibility through the fax-on-demand application. Callers with rotary phones access the application through voice recognition.

The company plans to add "SmartAnswer" a function that tells callers their queue position and wait time, and provides options for automated call back or transfer to voice mail. In the future, Wygant's Data-Follow-Me CTI application will let the caller's record appear on the agent's screen along with the call.

HIGH TECH
ACD

Things are changing in the customer service center of a well-known imaging company. They are handling more calls with fewer people. They are outsourcing more of their calls. They are using fax to help many of their customers faster. And they plan on using computer-telephony integration to serve their customers better.

A new OmniWorks telephone system from SRX is helping them with all these changes. The OmniWorks system integrates an ACD with IVR, voice messaging, fax and outbound calling functions on a single platform.

The company makes optical character recognition (OCR) and imaging technology. Their center has 20 agents handling customer service and supports five sales agents. Another five agents in another division of the company also uses the OmniWorks system.

One reason the company's customer service technologies manager appreciates the system is, as more of a computer guy then a telecom guy, he understands it.

"The system's open architecture makes it in some ways like a big PC," he says. "From a structural standpoint, it's easy to understand."

The system's flexibility and ease of use has really come in handy when dealing with outsourcing problems. Instead of waiting around for help if there are problems with the T-1 to the outsourcing company, the manager can rearrange things so the calls are handled within the center. "The calls are not stuck," he says.

That flexibility is also beneficial since the system is new. Call center management is finding it is changing things frequently as new features, functions and techniques, now made possible by the new system, are discovered.

The company is taking advantage of OmniWorks' integrated fax by giving callers the option to go to their fax-on-demand system, which has both sales and support documents — including the answers to commonly asked questions. The center is sending about 150 faxes a day.

"Every phone call we don't have to answer saves us money," notes the manager. "We like to automate as much as possible."

Toward that end, the company is starting to put OmniWorks' extensive announcements and prompts to work in an audiotext system. The manager said the company had such a system before, but it was difficult to modify. With OmniWorks' flexibility, he expects this system to be much easier to keep current.

The center is also using the systems' integrated messaging function, but not for customer support. the manager says it makes sense for the small sales department to use OmniWork's call-back messaging, because it helps them handle all their calls with a limited staff.

But the manager is hesitant to have the 20-agent support group use it. "Call backs are the least efficient way to handle calls." He feels the support group's size lets it handle all incoming calls without falling back on messaging.

The future for the company support center certainly looks integrated. the manager is looking forward to getting started with CTI. The OmniWorks' open architecture should make the task of integrating computer database to telephone system easy, he says.

3

COLLECTIONS

AGENCIES
PREDICTIVE DIALING

"The good news is the dialer can be customized to do anything you want it to do. The bad news is you have to decide what you want it to do," says the manager of a Midwestern collections agency.

Actually, it's all good news for the manager. With 10 years of experience in the field, he knows exactly what he wants his dialer to do. "I'm into customizing. This system has allowed me to do it."

He says the predictive dialing vendor considers its product to be software. This frees him to buy his own workstations. "I'm not locked in to buying a proprietary keyboard from them. They offer the service, but I don't need it."

The manager recognized early that collections with an automated dialer is completely different from manual collections. His company has two separate groups, each with about six collectors. One works with the dialer, the other works manually.

In the automated group are part-timers, hired specifically to work with the dialer when the company bought it over two years ago. Full-timers make the manual calls.

The strategy has been a success. The dialer, the manager says, has helped the company, "increase the number of contacts with debtors, which produces more payments, which means better recovery for our clients."

PREDICTIVE DIALING

How did one collection agency their calls before the bought a predictive dialing system? The technology was very versatile, but not very efficient. "We had the finger," says one staffer.

A conversation with the company's financial team reveals that the biggest value of the dialer is the added level of productivity and efficiency it has given to the operation.

"An agency needs to have a dialer work the way they work," says a staff member, "not the way someone else works." The collection agency found this in their predictive dialing system.

The team says most predictive dialers require a hot-key switchover to another software system. Usually this "other" software, the collections software in this case, is the most important. With this predictive dialing system the agents work with live data.

This is especially important to this company because they have one group of more than 40 "collectors," working manually and another group of 12 "agents" that work on the automated system. Each collector is assigned accounts, but agents make calls where needed.

The company uses the dialer to pick up the pace, where needed, and to meet goals. They may select to run all accounts from a particular client through the dialer. Or if an collector is on vacation, they may choose to have the agents work that collector's accounts until his or her return.

Agents leave messages with the assigned collector's name and direct telephone number. If they reach the debtor (make a right-party contact, in collections-speak) they talk to the debtor as if it was their own account.

The folks at collection agency feel the dialer helps them fulfill their mission because it fits the system that was already in place. It is a tool used by an "unselfish department" to help out all their collectors on an equal basis.

CABLE
PREDICTIVE DIALING

A cable company wanted a collections package combining hardware and software created specially for the needs of the cable industry. It found special software for Telecorp Systems' predictive dialer.

Using the system, it increased of collector productivity up to 300% and a three to nine month payback period.

The software keeps track of payments received daily, so the system only calls those who haven't paid. It also keeps track of customer's promises to pay and automatically calls them back if the payment is not received as promised.

CALL CENTER MANAGEMENT

As the customer service and collections manager for a Texas cable company, John finds the information he gets from his call center management software system valuable. But what he likes best about it is how accessible that information is.

The cable company's call center uses Telecorp Products' ACD Performance software to track call information, analyze agent performance and plan staff schedules.

John says he gets this information with a glance at the screen. "The information is color coded. If it is important, it flashes. You don't need a lot of training to understand it."

About 12 agents are on the phones in the cable company call center at any one time. They answer calls about billing, service problems and take orders.

The system tells John how many agents are actually available at any time. It also tells him how many calls were taken, how many were abandoned and how long callers wait before they hang up.

He can create reports to show each customer service representative how many calls she took and how many personal calls she made over any period.

The system also tells the customer service agents — using a LED readerboard display — how many people are waiting, the longest call waiting and the number of agents available.

Having all of this information makes the call center much more efficient at helping customers, says John. "A law says we must answer our telephones within 30 seconds. Using this system we can do that and know that it happened. The system tells us."

It also helps them save money. Deciding to add another agent is a much more exact science with the software package, John reports.

But it all comes back to how easy the system is to use. It is clear that it is this feature that makes the software system worth having in the cable company call center.

"I have some computer knowledge, but I would prefer not to use it every time I look at the screen," he says.

FINANCIAL SERVICES
PREDICTIVE DIALING

"If you can't measure it, you can't manage it," is the motto of one assistant vice president at one of the nation's largest credit card services companies. With this in mind, it is no wonder that one of the features the assistant VP likes best in the company's Melita predictive dialing system is the reporting capabilities.

But in this operation it is not just management that does the managing. The assistant VP especially likes the fact that the system lets collectors see their work statistics on a graph. The graph can show how many promises to pay they collected that day or other statistics. The immediate feedback gives the collectors control and makes their day go faster, says the assistant VP.

He is also impressed by Melita's cancel-dial feature, which gives collectors even more control over their own destiny. He says most predictive dialers base their pacing on the speed of a group of agents. If the best agent handles a call in one minute and the worst in four, most dialers will have the best agent drumming her fingers waiting for calls while the worst falls hopelessly behind.

The Melita system, he says, has an "extend time key." If the collector is not ready for the next call, pressing the key grants the collector extra time. This in turn causes the system to cancel calls before they are launched.

"Most other vendors populate from a hold queue. We consider that a nuisance," says the assistant VP. Using the cancel-dial feature lets his company eliminate that nuisance.

The department has 200 collection agents on the floor. Up to 99 of them can be on the dialer. All of the collectors working on 15- to 30-day debts use the predictive dialing system. The percentage of agents using the dialer goes down as the length of the debt goes up.

The assistant VP also likes the fact that his collectors can work at PC workstations running a Windows-based application. He feels limiting the keystrokes made by the collectors increases their productivity. He also likes the fact the collectors can now use the company's LAN-based computer system.

GOVERNMENT
PREDICTIVE DIALING

One state uses a revenue management system from DSI to collect delinquent state income taxes. It has succeeded in increasing the number of delinquent tax collection contacts made to 150%.

The system combines predictive dialing with the state's existing data systems. When a call is made, the collector knows the taxpayer's filing and payment history, and even his or her favorite excuses for not paying. Delinquent state income taxes add up to billions of dollars nationwide.

PREDICTIVE DIALING

A subsidiary of a regional Bell operating company wanted to allow access to the dialing systems to collectors at 140 geographically distributed workstations.

It used three predictive dialing systems from International Teleystems Corporation's (ITC).

The system's Smart Link gives collectors access to the dialing systems no matter where they are located. Also in the package is a feature which distributes either inbound or outbound calls to operators on a call-by-call basis.

UTILITY
ACD

Sophisticated call center technology plus an extensive training program equals great service for the customers of an electric company in the upper Midwest. The company recently won a customer service award from a utility industry association.

The utility has made big changes in its call center the past year, reports the team leader for communications technology. A brand new call center in one location handles all customer service and sales calls, while another call center handles both inbound and outbound telephone collections calls.

An Aspect CallCenter ACD routes calls at both locations, assisted by many network routing features (such as time-of-day routing) from the utility's toll-free service provider, AT&T.

A Queue Time Monitor from Croyle & Associates announces expected wait time to callers during busy periods.

The team leader aptly notes that in the electric company's call centers — as in most call centers in the utility industry — it is the plans for handling calls during an outage that drive the plans for call handling in the center in general.

The electric company is well-prepared to handle the influx of calls that invariably follow a service interruption. First, calls can be overflowed between the two call centers. In an outage situation, that means the collections staff of 80 or more can assist in answering outage report calls.

(At other times the 150 agents in the customer service center can handle overflow from the collections center if need be.)

Interactive voice response (IVR) also helps handle outage calls. A network IVR application from Syntellect lets customers report an outage by voice or using their touch-tone phones, says the team leader.

An IVR system from Periphonics also gathers outage information from callers, and does several other applications too. This system handles bill inquiries, lets customers switch to budget billing, provides a locator system for utility payment center, and even handles payment arrangements for customers who are behind in their bills.

A fax response system lets callers leave their telephone number to report an outage. Those numbers are sent to the utility every 15 minutes. The team leader says the telephone number gives the electric company the information it needs to help pinpoint the outage.

The electric company is well-prepared to handle customer calls, but its philosophy is not to sit back and wait for customers to call. One technology that helps them reach out to customers is a Melita predictive dialing system.

This system is used for making marketing calls in the customer service center. And through the utility's telecommunications network, the dialer's features are available to the collections agents in the other center too.

This policy of reaching out to customers is best seen in the company's customer service program. The heart of this program is extensive training.

The program's Process Manager, explains that in the electric company's soon to be deregulated market, the only way to stay on top is to not only be attentive to customers' needs, but to educate the customer — to be a consultant as much as you are a source of energy.

The electric company's training includes four weeks before the employee goes "on line," says the Process Manager. Two additional weeks weave this training with hands-on experience to give employees the "why" behind the company's policies.

The Process Manager notes that the training process tackles all parts of the customer interaction. Employees are not just given information on how to use the technology; they are trained in the entire business process, from the structure of the conversation with the customer, to the information content in that conversation.

So far the results have been stunning (and award-winning). The electric company consistently comes out on top in after-transaction surveys of sample customers, benchmarking comparisons and other in-house evaluations.

And there is more. "Last year we received over 90 unsolicited letters or calls thanking us for our service," says the Process Manager. When you consider what it would take to inspire you to write a thank you note to your favorite utility, it easy to see that this electric company has made a positive impression on its customers.

4

COMPUTER
TELEPHONE
INTEGRATION

BANKING

The life of an in-house telemarketer is not easy. Often you are limited to clients within your company, but your clients are free to choose an outside telemarketing agency.

That's the way it works at the telemarketing center for one of the nation's largest banks. To keep pace with even the most advanced outside agencies, this telemarketing center has tapped into some very sophisticated technology by installing a software and computer/ACD integration system from NPRI.

One of the advanced features the bank is using is simultaneous voice and screen transfer. The main benefit of this feature is the ability to transfer more difficult calls to specialists without requiring the specialist to start from square one with the customer, says the Executive Director of telemarketing division. "It allows us to provide the best possible service based on the needs of the customers."

The 180 sales reps in one call center and the additional 180 in another location mostly sell credit cards, but also sell an array of the bank's financial service products, says the director.

The telemarketing centers manage big volume fluctuations and provide job enrichment to sales agents by systematically flipping agents between inbound and outbound queues.

In the future, the bank hopes to have true "universal agents" that handle both inbound and outbound calls on an as-needed basis through the NPRI system.

The has used the NPRI system for about a year now, says the Director of Information Services. The implementation, he says, "ran surprisingly smooth."

But the telemarketing center can't rest on its laurels. They have to keep up with the latest call center technology advances. They plan to keep fine-tuning the NPRI system to more closely link the ACD and VAX computer so there is no redundancy in their reporting, and they hope to have their scripting language more on-line with the rest of the system.

These changes will assure the best service possible for both bank's internal and external customers.

CATALOG

A mail order hardware, plumbing and electrical products company doesn't take its customers for granted. The company knows that customers who are served quickly and efficiently will order again. It also knows that computer telephone integration (CTI) shaves time from calls and helps polish customer service performance.

But would implementing CTI mean junking its old systems? Not with the Aspect CallCenter ACD and the Aspect Application Bridge. With Application Bridge, the customer account information stored in an Oracle database on Sun host computers is available to the agent when the customer's call arrives.

This world-class call center handles about 8,000 inbound calls a day from its 160,000 customers with 144 agents.

HEALTH CARE

Imagine being rushed to the hospital in the middle of the night with abdominal pains. Quick! What's the name of your health insurance provider? What's your group ID number? Where is that handy insurance ID card?

Emergency medical care and careful record keeping don't exactly go hand-in-hand. A Midwestern hospital a computer-telephone link keeps both care and information flowing.

The hospital uses IBM's CallPath to link an IBM VRU, a Rolm 9751 and their computer system. The link makes things better for both patients and staff.

For example, the admissions department uses the integrated system to find patient records with incomplete information. "The system creates a worklist and helps direct calling activity," says the Senior VP of Information Systems at the hospital.

Patients can get information from the hospital's financial records department 24-hours a day through an IVR system. Patients enter an ID number and choose the information they want from a menu of choices. Easy questions such as account balances and last insurance payment are answered by the IVR system. Trickier questions go to a live rep, along with a computer screen on the patient's account.

The integrated system also helped put an end to the double booking of doctors' appointments in the hospital's clinics. These appointments are often made months in advance and once it was common for patients not to show up, wasting the valuable and expensive time of doctors whose services could have been used elsewhere. To assure work for the doctors, receptionists used trained guesses to book appointments. Sometimes they assigned two patients to one doctor at one time.

Today the hospital uses an automated dialing system that calls patients three days ahead of their appointment and confirms their attendance. If the patient indicates the appointment will not be kept, the call is transferred immediately to a scheduler who makes a new appointment for the patient. Newly opened slots are filled by other patients.

A similar outbound dialing application organizes work in the hospital's collections department.

The hospital has been using CallPath since late in 1991. But these advanced applications didn't spring up overnight. They are part of a continuing process for the hospital. "We are always looking for opportunities," says the VP. In the near future a time and attendance data collection system will combine badge-swipe information from employees paid by the hour with a telephone interface that can be used by salaried employees who don't swipe in and out.

PACKAGED GOODS

Everyone likes ice cream. That's good news for a large Canadian diary. They sell milk, yogurt, premium ice cream, various ice cream novelties and other diary products.

Come summer, the company's telephone lines are jammed with customers wanting to place ice cream orders. Sometimes the customer hangs up before a telemarketing representative takes their call.

A few years ago, that was a big problem, says a company rep. Back then the company didn't know anything about those customers who called but weren't connected.

Today, thanks to a system that includes an IBM AS/400 computer and a Northern Telecom Meridian 1 PBX tied together with a CallPath/400 link, abandoned calls are put in queue and the call is returned using ANI (automatic number identification) information.

The company reports that its customers, who include retail stores, food service providers, confectioneries (corner stores) and hospitals, have been universally delighted with their call-backs.

This delight means the company has already achieved one of its main goals in implementing the system: to improve customer satisfaction.

Also toward that end, it is collecting a database of customer telephone numbers to expand the reach of their ANI prompted screen pops. With the old system each account number was associated with only one telephone number.

So when a school cafeteria lady (a typical customer) called from home after a long day of serving sloppy joes at school, her call was delayed slightly while the rep called up the customer screen.

Now when that screen is called up, the rep asks the customer if he or she will be calling from that telephone number in the future. If the answer is yes, the number is added to the company's database to speed future calls.

The new system has meant other improvements as well. Once its 20 or so reps were split between inbound and outbound calls, with three or four handling inbound calls. Now all the reps make outbound calls on the system until they are needed to answer an inbound call.

As the reps dial through the customer base, a process automated through the AS/400, they can pause in the cycle of selected accounts, take a single, waiting inbound call then return to their outbound call, even after taking just one inbound call.

Another neat feature that the company is still working on is a "shopping list" that appears on the screen for each client. The list includes the top 20 products that the customer orders. It serves both as a reminder for the agent and as an upselling tool.

The company can add seasonal or other special items to the list. For example, in December "eggnog" might appear on every customer list to remind both rep and customer that it is time to stock up.

What is the ultimate sign of how this new telecommunications system has changed the way the dairy runs its call center? The company would like to upgrade job title of the company's telemarketing representatives to "inside sales representatives."

UTILITY

A utility wanted to serve 19 offices throughout the region with a single ACD while offering each location state-of-the-art technology. It used Teleoquent's Distributed Call Center ACD.

Using the system, the 19 offices, which used to be served by 19 ACDs, now work off a single switch, letting agents from other locations answer calls when one office is swamped with calls due to a major power outage. Automatic number identification (ANI) and computer-telephone integration (CTI) provide a screen pop for each call, giving the rep information about the customer's location, services and account status.

Dialed Number Identification Service (DNIS) and several different toll-free numbers help the utility automatically prioritize calls. A call coming in on the special toll-free number for gas leaks is answered ahead of general account or customer service calls that are not as urgent.

5

FUNDRAISING

OUTSOURCING

Sometimes it takes the worst of times to bring out the best in people. The hours, days and weeks after the Oklahoma City bombing saw heroics, not only from rescue personnel on the scene, but also from average Americans who donated money to aid the victims and their families.

An inbound and outbound telemarketing service bureau based in the Midwest helped make this generosity possible. Within one hour of receiving news of the tragedy, they had inbound agents standing by at all their inbound locations to receive calls for the Red Cross.

The first day the company handled as many as 40,000 calls with donations from the Red Cross, says the President of one of the company's divisions. The total call volume over the next few weeks was close to 100,000 calls.

How could the service bureau move so quickly when even a speedy service bureau takes a week to a month to set up a campaign? The division president and VP marketing give these reasons:

- A long-standing relationship with the Red Cross. The company has been working with the Red Cross for seven years.
- Systems and scripts were already in place. Minor adjustments were all that were required to tailor existing scripts to this disaster.
- Specialized training was minimal, since the service bureau had handled other disaster campaigns for the Red Cross and their agents

were experienced in handling this unique type of call. Pre-shift meetings explained the situation to agents.

- Previous experience with other disasters that didn't hit as quickly gave time for planning and adjustments.
- Ability to use agent positions across all their sites. The service bureau had 5,500 workstations in 14 call canters. They could quickly allocate these resources to meet the Red Cross's very high call spike.

DIALING

The folks at a blood center in the Pacific northwest have used a Mosaix predictive dialer for nearly two years, but the memories of manual dialing are still fresh.

"With Mosaix, we have been able to surpass by five or six times the number of calls we could make manually — while reducing the number of hours we call," says the Supervisor, Donor Services for the blood center. "The ability to manage those calls is unbelievable. We've been able to talk to a tremendous amount of people and share our message, but also be selective about who we reach."

The blood center provides blood from volunteer donors to 13 counties in Washington state. The 11 stations on their Mosaix dialer operate eight hours a day, five days a week.

A mainframe computer contains a database with not only donors' names, addresses and telephone numbers, but also important information like blood type and the date they last donated blood. (You should not give blood within 56 days of your last donation. The database keeps track of this and filters out donors who have given in the last 56 days.)

This information lets the blood center pinpoint donors that meet the region's current needs. The center can target people with a particular blood type in a particular town, for example. Or they can cast a wide net for all eligible donors.

Lists are built in the mainframe, then downloaded into the predictive dialer. At the end of the day updated records are uploaded into the mainframe from the dialer. The supervisor sees the dialer as a mechanism to increase calling, and as a filter that translates information from the telerecruiter on one end and the computer at the other.

The dialer has also let them extend their reach beyond existing donors. With the dialer in action they have begun to purchase outside lists for cold calling — a task that is much easier and much more productive with the automated benefits of a predictive dialer.

DIALING

Sometimes a predictive dialer means more than the ability to make a lot of calls in a short period of time. On a Native American reservation in South Dakota, a 16-station EIS 2400 dialer also meant jobs in an area with 80% to 90% unemployment, says the director of a local native heritage organization.

"We went into business to build a Native American heritage and cultural center in the Black Hills of South Dakota, on the highway going to Mount Rushmore and the Crazy Horse Memorial," says the director.

The director, one of the few non-Native Americans in the association, can count his experience in non-profit organizations, data processing and telecommunication in decades, not years. When the association's first fundraising call center was started he went with EIS's 9600 predictive dialer based on past experience.

"The EIS system, quite frankly, is easy to install, set up, to use and maintain. And EIS has a pretty good service department. The reservation is exactly in the middle of no where, but when there is a problem, they get there in a day or so," he says.

When the association expanded their call center to the reservation itself, its goal wasn't so much fundraising as to create jobs for about 20 people. The director had hoped to expand the system from 16 stations to 24 stations next year. With the expanded system, the association wanted raise funds to build a warehouse on the reservation to stockpile food and clothing for distribution to needy tribe members.

The original expansion was self-sustaining in 30 to 60 days.

"To my knowledge," says the director, "this is the only telemarketing installation on an Indian reservation anywhere." But in spite of the center's success, and its goal to increase employment on the reservation, the association has no plans to offer telemarketing or fundraising services outside the association's needs.

The director says the legal complications on the federal and state level would make that too difficult. The number of "in-house" projects could keep many more agents busy, though.

Sadly, the last time we spoke with the director, the reservation telemarketing program had been closed down indefinitely. It was a good idea, and we hope its time will come again.

DIALING

A fundraising service bureau based in Houston wanted to lessen hiring chores by reducing staff, while increasing or maintaining revenue levels. It used a predictive dialer and integrated outbound autodialer, both from Stratasoft.

After implementing these products the service bureau was able to reduce their staff by over 50% and increase sales by 200% to 400%. The agents tripled their efficiency and employee satisfaction is at an all time high. The predictive dialer allowed the company to automate pledge callbacks. The system contacts 1,600 pledgers an hour with a thank you message and a notice to those who have pledged and not paid. The system has increased pay-ups.

DIALING

A fundraising service bureau wanted to improve the efficiency of lists, which they feel have deteriorated over the years due to people moving more frequently. "A campaign that used to take a month or two a few years ago now takes three to four months."

The service bureau uses a predictive dialing system. The dialing system is a client/server system that runs on the Novell NetWare platform. It has many list management features and works with or without a PBX interface.

Using the system, the work done by ten people in ten weeks with a 30% ratio of live voice answers now takes six predictive dialing stations two to two-and-a-half weeks with a 65% ratio of live voice answers with the predictive dialing system.

DIALING

Diversity is the name of the game at another Midwestern service bureau. The company started out with the majority of their work in fundraising. Today, fundraising is still one of their strengths, but they've moved on to sales campaigns and marketing surveys too.

Even within the realm of fundraising, the company is diverse. Not only do they do consumer fundraising, but they also have expertise in large gift fundraising, where the starting pledge is $50,000.

The telemarketing software and dialing system that keeps up with all this variety is the Telescript system from Digisoft Computers. "We started with 10 stations of Telescript and kept adding and adding and adding," says the company's president and CEO.

The company has a total of 300 stations, 60 inbound and 240 outbound. They have grown their Telescript system to 125 stations — so far. The president says future plans call for 90% of the company's stations to be on the Telescript system.

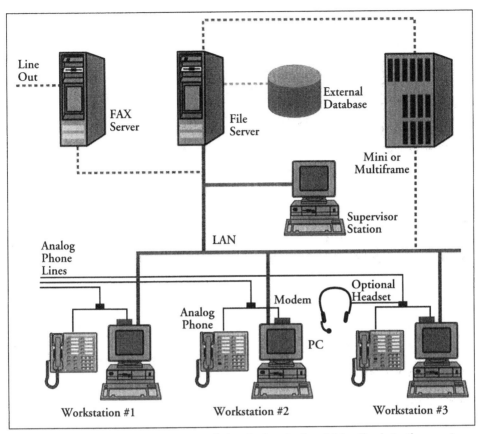

Line Out

FAX Server

File Server

External Database

Mini or Multiframe

Supervisor Station

LAN

Analog Phone Lines

Optional Headset

Modem

Analog Phone

PC

Workstation #1

Workstation #2

Workstation #3

This is a typical Telescript set up. It is similar, but not identical, to the system used by the fundraising service bureau. Integration and programmability make it flexible enough to keep up with service bureau's diverse needs.

Not surprisingly, one of the things he likes best about the system is that it provides a boilerplate with which you can build the custom system you want. The service bureau has gotten so good at creating custom applications with the Digisoft system that they now provide custom programming for other companies that don't have an in-house programmer.

The president also likes the variety of automated dialing methods available. For consumer fundraising predictive dialing is the key, and Telescript provides it. For their large gift fundraising campaigns, where they are often trying to reach CEOs at work, the agent generates the call by hitting a key that sends out a dial string.

It's also possible for the Digisoft dialer to set a pace for the agents, says the prez. The dialer can be set for an interval, perhaps four seconds if no call wrap-up time is required. If the agent doesn't generate a call after that time, the call is dialed for them.

The president also praises Telescript's reporting system. His favorite screen has bar graphs that get longer as the agents spends more time on the phone. Because he knows how long an agent will probably spend on a given call (a particular survey has been calculated to take ten minutes, for example) he knows when that agent will be available for training, a break or a conversation.

Whether you are asking John Q. Public for a few dollars or a CEO for thousands (the company once got a $500,000 pledge over the phone), the agent's skill set is the same, says the president. For the large gift campaigns he looks for agents that are super articulate, have perfect diction, think on their feet and who can ask for a few thousand dollars like they are asking someone to pass the butter.

The trick to these big ticket campaigns is getting past the gate-keeper, he says. One way to do it is to have a prestigious name, such as a well-known politician. One campaign the company runs involves a call that alerts the potential donor that a video tape from a prestigious person is coming, the mailed video, and a follow up call that asks for a donation.

LOOK-UP

Directory assistance is not what it used to be. The Manager of Information Services at a well-known Boston college remembers when you could request two numbers per call. Now the students in her office get only one number per call, often wait longer to get the number, and pay between 43 cents to 75 cents per number.

In spite of this, the college is updating their alumni records even more effectively now. The solution is the Electronic White Pages from DirectoryNet (DNI).

The college uses the Electronic White Pages to add a telephone number to a name already in the college's records. Various departments at the college call alumni for fundraising, to follow up on mailings announcing special events for alumni, or for other reasons.

Students, working on Macintosh computers, determine which records need telephone numbers during the day. At night (it's cheaper), those records are sent to DirectoryNet's IBM compatible computer system by modem. The appended records are sent back the next day.

The Macintosh-IBM link took a while to work smoothly, but the manager is pleased with the results.

Even though the process takes two days, it is still faster than having the students call directory assistance for each number, says the manager. And because of the volume of requests the college has, the bulk rate for the information comes to just pennies per number — much cheaper than it would through directory assistance.

Last but not least, she notes that because the students waste less time dialing directory assistance, they have more time to do other work.

APP DEV SOFTWARE

An Ivy League university wanted to centralize donor information maintained separately by the university's ten schools to successfully support a campaign to raise $2 billion in five years.

The university developed the client/server-based fundraising management application with Blyth Software's application development software. The development software, and the application, work in both Macintosh and Windows environments.

Using the app development software, team of about eight people developed and implemented the software in just eight months. Fundraisers are able to track, target and report on prospect alumni on a University-wide basis. Within one year of implementing the system, they have raised 50% of their five-year goal.

The software system has over 200 users and has tracked and managed nearly 600,000 prospects.

6

HELP DESK

BANKING
HELP DESK SOFTWARE

The support services division of a large bank wanted to centralize help desk support in a fast-growing organization. It used Software Artistry's Expert Advisor, running on a Novell LAN

The support services center has 27 analysts. The center handles 15,500 to 17,000 calls per month.

The bank plans to use the knowledge bases in Expert Advisor so they don't have to be so dependent on specialized knowledge among analysts. They expect to be able to close about 95% of their problems without transferring them to another department.

OUTSOURCING

A large, Manhattan-based bank had all of its help desk bases covered between its information services department, an outsourced help desk service and various hardware repair services. However, they wanted to make things easier for the 7,000 employees that use these help desk services.

The bank contracted with The Sutherland Group, a help desk service company, to provide a single point of contact for its employees. Hiring Sutherland the bank kept its on-site IS support and its relationship with several hardware support companies while giving its employees the benefits of dialing just one extension for all their support questions. Sutherland dispatches the proper service and follows up to make sure the problem is solved.

GOVERMENT
PROBLEM MANAGEMENT

When the world is watching, you have to be on your game. That's why the a state transportation department started using Silvon's Helpline help desk software to handle problems with its computer systems before a major international sporting event.

The department's computer systems have over 2,000 users in the region's largest city and around the state. The department needed to find a reliable way to log and track work requests when there were problems with its local or wide area networks.

Helpline, an enterprise-wide help desk system with automated problem logging and assignment, filled the bill, letting the department log and track requests, and generated detailed multiple reports for management immediately.

Helpline also has built-in problem research agents, automatic problem escalations, status notifications to end users, and interfaces to e-mail and the Internet. While the sporting event was not glitch free, at least the transportation department was prepared for everything the world threw at it. And it continues to use the Helpline system.

TRAVEL
PROBLEM MANAGEMENT SOFTWARE

When the Information Service Department of a major hotel chain decided to bring support for their proprietary property management system in-house, they decided to do it correctly from the start. Part of doing it right meant having a problem management software package in place from day one.

"I had heard that starting with pen and paper only makes the conversion harder," says the hotel's Director of Customer Service. She says problem management software from Quintus was chosen because it was easy to set up a database of internal customers, easy to find problem records using simple search parameters, and easy to document solutions to common problems for training and reference.

Also important is the system's relational database design. The help desk plans to build a problem-solving knowledge base as they go along.

The fact that the system is Windows-based lets staffers cut and paste from CC:Mail and other applications — helping to build the database quickly. The report writer and query function lets the help desk work with the property management systems' development people to answer questions like, "How often has this bug surfaced in the last 60 days?"

"Having a problem management system from the beginning has helped us quickly gain credibility with our customers," says the customer service director. "Our customers know that we know what we are doing and have everything under control."

As they expand the help desk to support more products, having the problem management system and the knowledge base will allow them to hire generalists, not specialists in specific products. "The only way to do that is to have knowledge in the system," she says.

UTILITIES
HELP DESK SOFTWARE

You take it for granted: turn on the faucet and clear, cool water rushes out. In one large city in the mountain West, you can thank Applix help desk software, in small part, for this.

How does help desk software bring water to people's homes and businesses? It works like this: the help desk supports the computer hardware and software of the people who maintain the reservoirs, treat the water, run the recreational facilities, and maintain the pipes and water mains. Without computers, the rest doesn't happen.

This western city growing by leaps and bounds. The water company keeps up with this growth through a commitment to service excellence. This commitment is found even in the company's help desk, which has made a service level commitment to the people that they serve.

The Applix software plays an important role in keeping this commitment, says the department's Application Developer. "Once a request is logged, the work status is tracked to make sure the customer's demand is met according to the service level agreement," says the application developer. Without tracking the help desk's response with the software, it is possible for customers to fall between the cracks, she says.

The 14 staffers in water company's Operation Support section of its Information Services department work hard to make sure customer requests are filled as soon as possible. In urgent cases, this can mean a promise of a solution within 24 hours.

The help desk handles an average of 24 calls a day and 6,300 calls a year.

Not only does Applix Enterprise track problems, it also gives Operations Support people the information they need to solve problems faster. the application developer says, "It's easy to store things away, but when it comes time to retrieve the exact thing you need, you need something like Applix Enterprise to make the process easy and user-friendly."

As people and businesses continue to flock to the city, the surrounding county and the 76 communities that have contracts with the water company, the demands on the company's help desk will continue to grow. But with help desk software to retrieve information and track the level of service offered, it looks like the water company's Operation Support is well-equipped to handle those demands.

READERBOARD

The buck stops at the Customer Service Desk of one national telephone service provider. Until recently, over 50 other points of contact within that company also provided systems support. A consolidation was in order.

The Help Desk Consolidation project means not only saving money by eliminating redundancy, but it's also a chance to do things a little better. One of the things this help desk wanted to do a little better is communicate quickly with all agents and administrative staff.

After a rigorous selection process, Texas Digital Systems was selected to install a Quickcom readerboard system in the help desk's new location. This system integrates not only with the help desk's CTI-based ACD, but also with its trouble management system, Action Request System (ARS) from Remedy.

Since this past fall, the new help desk has been up and running, using the TDS displays. The displays include information about different queues and are wowing visitors with their state of the art technology.

7

INFORMATION SERVICES

BANKING
IVR

This bank is big. Their five customer service centers throughout the south handle over six million calls a month.

Each center uses sophisticated call center technology to serve callers with both IVR information and support from live agents.

Managing the development and implementation of new applications and technologies in so many centers is not easy, but the company has a clear goal of what they want each center to do.

The bank has created a call center in Richmond, VA that serves as a model for the other centers. "The Richmond VRU script was developed as the bank's model script to provide customers access to their account information quickly and easily any time," says a bank vice president.

The Richmond center uses a Rockwell ACD; Early, Cloud customer contact software and an IVR system from InterVoice.

The VP says a major factor in selecting the IVR system was InterVoice's experience integrating with the other vendors' systems.

That integration is very important since the bank uses computer- telephone integration (CTI) to deliver customer information to their customer service representatives.

All calls to the center are answered by the voice response system.

Callers are given the choice to use the IVR system or to speak with a CSR. All callers are asked to enter identification and account information. When the call is transferred to a CSR, a screen of customer information is transferred with the call.

This is particularly helpful when a caller runs into trouble and bails out of the IVR system. The CSR knows where the caller was in the menu structure, which often tells the CSR what area of information the customer is interested in, says the VP.

The end result is customers spend less time on the telephone and the CSR can help them more quickly.

Don't think the bank has forgotten the human element though. The VP gives much credit for the success of the system to the quality of the IVR script.

"A common problem is for people to feel confused or lost when using IVR. Our system has special features that always let callers return to a main menu. Callers can also cancel input by using the asterisk key — instead of starting from scratch," she says.

She also likes that the system is built to enable call center management to adjust their script to handle special situations — without help from technical support.

Evidence of the system's success is that 77% of calls to the customer service centers are handled by the IVR system alone. At the target Richmond center the average is 80%.

In the future the bank plans to keep the other centers moving toward the target, add even more IVR applications and launch Spanish and hearing-impaired versions of their current applications.

IVR

A lending company wanted to provide access to mortgage data quickly and easily, even thought the number of loans serviced had doubled.

It used the VMX Integrated Voice Processing platform with VMXworks.

This finance company was using three different systems to process calls, take messages and retrieve account information. The VMX system provided a single platform so callers did not have to be transferred from system to system.

ACD

A New Jersey bank wanted to upgrade a discontinued ACD with more lines and stations. It also wanted to integrate with an automated attendant and add new call processing capabilities.

It did this with an Applied Voice Technology PC-based ACD.

Using the system, callers are now directed to an IVR system for some information. If all 12 agents are busy they may choose to hold or leave a message. Agents see call status and other information on their proprietary agent telephones.

The number of calls handled has increased to about 1,000 per day and reporting has improved. Calls are routed between the IVR and a live agent and overall call handling is more efficient.

IVR

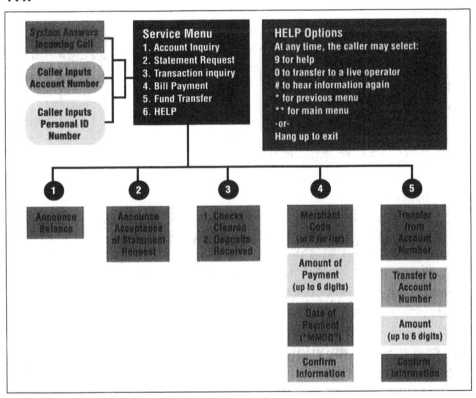

When you say "bank by phone," you mean bank through an interactive voice response (IVR) system. IVR systems are one of the most popular call center technologies used by the banking and financial services industry.

Periphonics was one of the first companies to put together an IVR banking application, and they are still at it. You can see their expertise reflected in the simplicity of this flowchart of a typical "bank by phone" application. Of note are the help options available at any point in the system.

ACD

A bank card company wanted to provide "one stop shopping" for all bank card questions. It used an Aspect CallCenter ACD, IBM imaging system using OS/2 Windows-based applications.

This 300-agent call center answers questions about the bank's ATM, debit and credit cards. The main center is in Virginia and is networked to a smaller call center in North Carolina through the ACD system.

FAX

A South American bank wanted to let customers request their bank balance or a statement of their last 12 activities. It used the Ibex FactsLine.

This is a Windows application in which and interactive voice response system and module host interface are connected through a PC gateway to the bank's Unisys computer system.

SPEECH RECOGNITION

A bank wanted to replace complex touch tone menus with speech recognition and save callers from waiting while long prompts were read.

They used a continuous, independent, large-vocabulary recognition system from Integrated Speech Solutions (ISS) and ISS' natural language processor.

The system lets callers speak their way through common banking IVR applications such as retrieving account balances, interest rate quotes, mutual fund quotes and making account transfers. The system comes with pre- trained home banking neural networks and pre-recorded prompts, but can be customized to fit a broad range of functions and telecommunications systems.

CLASSIFIED ADVERTISING
SPEECH RECOGNITION

A company that publishes a number of weekly classified magazines featuring cars, real estate and other specialty products wanted to provide superior customer service to advertisers and the public to show their commitment in a competitive marketplace.

Using a voice recognition and voice mail system, the publisher receives many calls for information on advertising, deadlines and rates. The system helps them provide a high level of customer service with its friendly messages and options to leave a message or arrange a callback when callers have to hold for too long. The system also provides callers with information on their number in queue and average wait time. This keeps callers on the line and helps them feel coddled.

SPEECH RECOGNITION

An IVR classified ad maker wanted to make the system easier to use for both buyers and sellers through the use of speech recognition. For example, with speech recognition, callers don't have to spell out on their telephone keypad that they are looking for an O-L-D-S-M-O-B-I-L-E, they can simply speak the make of the car.

They used speech recognition technology from Voice Processing Corporation.

With speech recognition in place, callers to the system choose to speak in English or Spanish. They say the make and model of the car they are looking for, the major features they want and the highest price they will pay. Callers say their zip code in a continuous stream. The system uses the zip code information to give callers the nearest car that fills their requirements first.

Sellers can modify their ads by speaking the new information into the system.

FINANCIAL SERVICES
ADDRESS CORRECTION

A major mutual fund company wanted to increase the percentage of correct addresses entered for new customers or customers requesting an address change.

It used address-checking software from LPT, a Pitney Bowes company. This software provides a lookup against a database of valid addresses. It interfaces with other applications (such as sales software) and can be run in real-time or batch mode.

Using the system, the mutual fund company is able to correct 10% of the new addresses. When you consider the company handles over 30,000 calls each day, that's a lot of wrong addresses. Assuming customers providing incorrect addresses would be lost to the company and each customer will invest a significant amount of money with the company the savings from using the system are huge.

IVR

A major brokerage firm wanted to enhance the services of its automatcd system which lets customers trade stocks and get quotes from any touch tone phone.

To do this, it used eight VAS/VPS 9500 voice processing nodes, VAS 4000 Voice Application Server, VRNA 2000 Voice Response Network Administrator, all from Periphonics.

The system handles over one million per month, with a growth rate of 15%. It sports 1,700 digital telephone ports. Callers get a 10% discount on commissions when they use the service.

HIGH TECH
HELP DESK SOFTWARE

A high tech company specializing in hardware for the graphics industry has 22 agents that handle up to 800 calls each day.

The company wanted to make returns for products under warranty more efficient and customer friendly. To stop duplicating efforts between two support departments: Technical Support and Service Fulfillment.

To do this, it used a technical support software system from Clarify.

Using the system, a complete record of the call entered by Technical Support, including information such as product failure code, is available to Service Fulfillment when Technical Support transfers the call to them for a defective product return. The call records are automatically updated by future actions, including the assignment of a Return Merchandise Authorization number and the recorded receipt of the defective product.

HEALTH CARE
ACD

A 822-bed, not-for-profit hospital in the Midwest uses computer-telephone integration to help its five-agent Patient Financial Services department.

An IVR unit answers calls to Patient Financial Services. It collects the callers' account numbers and provides routine information such as account balances. If necessary, the call is transferred to the ACD function and answered by a live agent. IBM's CallPath delivers patient information along with the call, so the agents already have the patient's name, account number and other info.

The hospital benefits from the technology because calls that used to take at least five minutes now take about two. The hospital's telecom manager credits the automatic retrieval of patient records for the improvement. Also, the number of agents needed to answer calls in the department has been reduced from seven to five. The other two now handle other duties.

RETAIL
ACD

At one of the country's largest office supply stores, you often wind up with more for less money. The retailer itself got more for less when it installed a Rockwell Spectrum ACD to handle store location calls and calls for its direct sales catalog.

At first the retailer thought using computer-telephone interface applications through the Rockwell Transaction Link for outbound calling and screen pops would be a solution for "someday," but once they found out how much agent productivity they could gain, they decided to implement the applications sooner rather than later. Similarly, implementing a voice/fax response system was less expensive than they thought it would be.

The system also includes Intervoice interactive voice response (IVR) and TCS management software.

Call center traffic was slowly shifted from the old call center in Massachusetts to the new center in Kentucky. By December the Kentucky center should be nearly fully staffed for its 440 agents positions.

TRAVEL
AUDIOTEX

One of the largest travel agencies in Germany, wanted to offer customers big savings on otherwise vacant airline seats and hotel rooms.

Trips have to be booked at the last minute — from a week to just a few hours before departure. Pulling these trips together means customers need information quickly, but having enough agents on the telephones was costly, and the agency didn't make a lot of money on the discounted trips in the first place.

The solution was an audiotext system from Moscom. Actually, the solution was 20 systems, each in one of agency's 30 discount offices.

Agents use a microphone to record new announcements about available travel packages. Callers select from "Last Minute," "Flight Only," and "Travel Planner" message categories, then hear information about packages in those categories.

When the caller hears something intriguing, he or she is transferred to an agent who books the trip. The system plays an excerpt from the selected announcement to the agent before connecting the caller.

The agent can answer, "Hello, I understand you're interested in reservations for our Zurich special..."

When no agent is available, the system records a message and lights an agent's message-waiting light.

The really interesting thing about this application is that, since there is no touch tone service in Germany, it uses voice recognition. Callers select the Last Minute category by saying "Last Minute." They transfer to an agent by saying "agent."

Callers may also ask to repeat an announcement, hear the next announcement or switch to a different category.

The agency's customers seem to enjoy using the system. Some call several times a week to keep on top of the latest discount travel opportunities.

8

THE INTERNET

HIGH TECH
HELP DESK SOFTWARE

Don't you hate it when you're are playing a video game in the wee hours of the morning and something goes wrong? Well, we won't name names, but for some people, especially the customers of maker of software game products, this can be a major bummer.

Enter Inference's CasePoint WebServer problem resolution software system. Now gameplayers get 24-hour-a-day assistance over the Internet through a self-help technical support tool that lets gameplayers use the same tools as the company's crack internal product support reps.

Gameplayers using the automated help desk are guided through a series of questions by a popular movie character. These questions are, of course, posed behind the scenes by Inference's CasePoint WebServer, and lead the gameplayer to a likely solution to his or her problem.

The automated help desk does not replace the company's highly qualified phone reps, but only complements them by adding to the hours and days help is available to the company's customers.

HELP DESK SOFTWARE

How can a company keep up with the frenzy of the Internet? For the

Internet subsidiary of a Regional Bell Operating company, it is with the help of Vantive HelpDesk software from Vantive.

The company introduced dial-up Internet access last spring, and knowing that success had killed other Internet service providers (ISPs) who couldn't keep up with demands for technical support, it knew it had to provide world class service to the flood of new subscribers.

Customization of its help desk software was the key to accomplishing this. In addition to responding to a standard Request for Proposal, Vantive had to pass a live customization test to win the company's business.

Today the company uses Vantive HelpDesk and an enterprise-wide tool that manages customer information, routes inbound e-mail (for trouble tickets) and uses electronic workflow to track the entire service delivery process.

Next, the company plans to use Vantive to help launch its dial-up ISDN service.

HELP DESK SOFTWARE

A provider of RAID and tape storage systems for LANs wanted to improve customer support by making solutions to common problems available to customers 24-hours a day through the Internet.

It used Apriori GT from AnswerSoft. The company paid off their investment in the Apriori system in less than a year. They saw a productivity gain worth $142,000 annually. And because most callers get answers on their first call, the company saves $70,000 a year in long-distance charges by not having to make return phone calls.

CALL TRACKING SOFTWARE

When you have 40 people answering 9,000 customer calls each month, it can be difficult to make sure that not one of those calls falls between the cracks. At a company that makes school district and local government software, every customer and every call receives the attention it deserves thanks to Service Call Management (SCM) software from ProAmerica.

After 10 years of using a system developed in-house, the 27-year- old software company turned to ProAmerica for a new system that would organize its various customer support groups, improve departmental time management and increase customer satisfaction.

With SCM, software company logs each call, escalates it as needed, and adds information to its knowledge base. The company finds it responds to new issues faster and has a permanent record of issues that have been resolved. software company likes how ProAmerica keeps on schedule with innovations, such as SCM Web, which lets customers get to SCM through the Internet.

9
KIOSKS

BANKING
VIDEOCONFERENCING

A company wanted to let customers at banking kiosks (mostly rural customers without a local branch) to do all types of banking, including opening accounts and obtaining credit.

It used a PictureTel videoconferencing system to achieve this goal.

This systems differs from video-equipped ATMs in that it lets customers meet with service reps, sign documents on the spot, get copies and return originals. Customers speak to the service rep through a handset.

RETAIL
HELP DESK SOFTWARE

This is the busy person's dream company. They make interactive electronic kiosks that (after a deposit of a credit card or cash) print out a gift certificate good at any one of a long list of stores you select from.

Providing technical support for these unstaffed kiosks is a challenge — a challenge the kiosk company is meeting with Top of Mind help desk software from The Molloy Group.

The unconventional capabilities that make Top of Mind ideal for the company's mission include the ability of the system to "learn" from problems that have already been solved, using a hybrid of several forms of artificial intelligence, including neural networks and fuzzy logic, as well as conventional text searching and parsing.

The kiosk company's help desk handles 1,500 trouble tickets a week. It is staffed by a lead technician, three senior and 12 assistant technicians, and runs 24-hours-a-day, seven days a week.

The support center is rapidly building a powerful knowledge repository in the system on kiosk and network problems, diagnoses and resolutions. Since the technology and its implementation are novel, there is no alternative source for this knowledge. Top Of Mind is absorbing the support center staff's cumulative experience and enabling everyone who supports the kiosks to share that knowledge. The system truly grows "smarter" every day.

10

MONITORING AND AGENT EVALUATION

CATALOG

The call center of one of the world's largest outdoor recreation equipment and clothing catalog companies operates 24 hours a day, seven days a week, reports its Director of Telemarketing. The center uses an average of 360 agent positions and from 400 to 500 agents to staff those positions. At peak — around Christmas — they use a staff of 1,000.

Not too long ago the outdoor cataloger evaluated their agents by monitoring them live and recording their performance on audiotape. "It required lots of cassettes," says the director. Using the tapes was cumbersome and time-consuming. Using tapes meant not only that the calls had to be recorded and listened to, but the tapes themselves had to be managed.

The company was poised to hire a sixth person to monitor and evaluate their agent's performance. But instead of just hiring another person, they decided to re-evaluate the processes.

They wanted the processes to take less time. They also wanted to increase security. "Cassettes are very easy to transport," notes the director. And they wanted to make the taping process more flexible and automated.

They found their solution in Teknekron Infoswitch's AutoQuality, an intelligent monitoring system. AutoQuality runs on a '486 PC. The system schedules recording sessions and records them in digital format. Supervisors or evaluators can play back the recordings from their ACD station set or a remote telephone.

The standard system has four ports and 100 hours of voice storage, but is available in versions with up to 20 ports and 300 hours of storage.

The director says, "It's been very effective." Password protection on the recordings has increased security. Storing the recordings as computer files has made them easier to manage. The computerized scheduling means recordings can be made on schedule, without a staffer handling the process manually.

Evaluators find they can listen to the digitized records much faster than they could listen to the taped recordings because they are so much clearer. Because of this, their productivity has increased. From five evaluators, the outdoor cataloger is now down to three.

"Using AutoQuality has been a positive experience for us," says the director. "It has simplified the evaluation process."

HEALTH

The customer service representatives at an HMO system in the upper Midwest are conversant in up to 80 different benefit package designs. When people call with questions — either enrollees or health care providers — the CSRs have to get it right, especially now that price changes have made their market more competitive and customer service is a major differentiator between competing plans.

It's a tall order. The manager of customer service for the HMO, and his team of supervisors and quality information administrators (QIAs) fill that order by monitoring each of the center's 70 agents 20 times per month.

The customer service manager says monitoring is also used as a training tool so the CSRs can continuously improve their skills. Experienced reps, who have heard it all and done it all, need a lot of feedback to keep the learning curve on the upswing.

Before the HMO installed an AutoQuality! monitoring system from Teknekron Infoswitch, there was a price to that high level of care. Before the system, supervisors and QIAs had to catch calls live, from beginning to end. Out of fairness, no call not listened to completely could be used for evaluation. That meant any time a CSR asked a supervisor a question in the middle of a monitoring session, the whole session had to be scrapped.

Without taping, the evaluators could only work during the CSR's hours and if a CSR disputed an evaluation, there was nothing much that could be done to rectify differing perceptions.

Since the HMO installed AutoQuality! in October 1994, that situation has changed. The system schedules agents for monitoring, making logistics easier. Monitoring sessions are taped, so evaluators can simply start and stop the tape when interrupted, without having to scrap the session. With the sessions on tape, supervisors can evaluate CSRs outside of call center hours, giving them more time to interact with their staff during the business day.

The CSRs are pleased too. They know if there is any question about an evaluation, they can go over the tape with their evaluator. The center is even able to use the monitoring sessions as training tools, using a tape of a particularly good or particularly difficult call to reinforce mastered skills and teach new skills using the agent's actual work.

The automated monitoring system is letting the HMO test a self-auditing program with experienced CSRs who have consistently shown high quality work. These reps listen to their own calls on the system and review their completed audits with a supervisor.

The customer service manager says he has seen a 25% to 30% increase in productivity in his supervisors. The ratio of supervisors to CSRs has risen from 1:10 to 1:13 since they started using the system. When the center started looking at extending their hours, monitoring was a non-issue in the decision, says the customer service manager. "It won't be necessary to have someone here to audit calls. We can just set up AutoQuality."

The HMO is in the midst of implementing Teknekron's P&Q Review software, which will further automate their monitoring and review process. CSRs helped design the review form and scoring system. The new system should help them improve even further the high quality service their customers have come to expect.

ORDER HANDLING

CABLE
IVR/ANI/MESSAGES ON HOLD

A cable programming provider wanted to increase pay-per-view buys among their affiliates' customers. It used Telecorp Systems' Home Ticket Intelligent ANI service with "Hot Spots."

Home Ticket is an interactive voice response system (IVR) which uses automatic number identification (ANI) to provide instant and automatic pay-per-view ordering. Hot Spots are short promotional messages played during the ordering transaction.

Using the system, A 30% increase in pay-per-view buys. One pay-per-view manager at an affiliate cable company noted the promotional messages "are particularly effective for the frequent pay-per-view customer."

CATALOG
ACD

When the first cry of "Play ball!" echoes across baseball and softball fields each spring, the folks at this sporting goods catalog company can breath a sigh of relief. Their busy season, which started at the end of January, is over.

It is a catalog that provides baseball and softball equipment to high schools and colleges.

Coaches plan ahead for the spring season and start calling the company in November. By February, the company has hit its stride, handling a peak number calls. By late April or early May (depending on the weather) the calls start to drop off as teams stop ordering and start playing ball.

The sporting goods catalog uses an ACD system from Applied Voice Technologies (the division that was formerly Telcom Technologies). The company's leader knows his call center inside and out. He doesn't need a report to tell him the peak season in this very seasonal business.

"The total users on the system numbers about 16 at any one time," he says of his small center. The ACD mostly serves to smooth out the wrinkles in call volume he says. "We are doing more with fewer people. And the phone is not always ringing in your ear."

The sporting goods catalog is making good use of the system's automated attendant and integrated voice messaging features.

Part of the voice messaging system is used for messages-on-hold. Those messages are "short and sweet" and created in-house says the boss. Voice messaging also relieve agents of repetitive questions, such as "What are your hours?"

With the automated attendant feature, the manager likes the fact that he can program in holidays and business hours a year in advance and not have to worry about changing messages or routing schemes before a holiday or at the end of the day.

He also likes the fact that the system can grow along with the company.

You might think baseball and softball players make the best agent, but the manager says they sometimes start talking shop and stop taking orders.

"We try to hire moms," he says. The flexibility of hours works for both the company and the working mothers.

It seems that for players, coaches and moms, the sporting goods catalog's Applied Voice ACD is a hit.

ACD

This call center is unusual, if not unique. It offers a variety of services to catalogers and direct marketers. These services include designing marketing programs, research, training for call center managers and representatives, and its 250-seat call center, which is open 24 hours a day, seven days a week.

Catalog clients contract with the call center in a variety of ways, says the company's VP of Information Services. Some contract for a certain number of seats, others for a consistent volume of calls, others off-load some of their volume during peak periods. The center offers both inbound and outbound services.

At the hub of all this activity is a Spectrum ACD from Rockwell. The Spectrum generates the center's outbound calls in addition to routing the inbound calls.

For the majority of its clients, says the VP, the outsourcer has direct access to their order entry system. This means when an agent calls up an order screen, he or she does not get the same, standard screen every time, but a screen that is actually being fed from the customer site. The big benefit to clients is that the orders are posted to their systems in real time.

To do this requires some high-powered technology — and some high-powered training. It's no surprise then that the outsourcer's own call center serves as a lab for its training techniques, and that techniques learned in their high-powered call center in turn influence what they teach others.

An agent at the outsourcer may work on as many as three different ordering systems. The company accomplishes this through the use of "shared groups."

The VP says that the Spectrum's routing schemes, including skills-based routing, allow the center to run this way. Other ACDs might put a call in queue for the primary group for a certain length of time, then send to the overflow group. But if an agent in the primary group isn't available, the Spectrum puts calls "into one big queue and flows the calls to whichever one comes first," he says. "There is a logic layer above the physical gate."

Skills-based routing means the ACD sends calls first to the agents with the most training in a particular customer's ordering system. Those with less training would be second choice. The outsourcer is also working on adding performance-based routing to the system.

Customers are identified through dialed number identification service (DNIS). Calls may come in on any carrier used by the customer. Needless to say, computer-telephone integration plays a big role in the center.

The VP reports that the outsourcer's sophisticated needs pushed the envelope of Rockwell's application development functions. "We found out just how flexible and reliable this system really is," he says. As a further testament to the system's reliability, he notes that there have been just six minutes of unplanned downtime in the two and a half years the company has used the system.

Plans for future growth are another reason the outsourcer has stuck with their Spectrum. The VP says the outsourcer doesn't plan to be a one call center operation forever, and the Spectrum will let them manage multiple call centers effectively from one site.

ACD

Back in 1986 one major TV shopping network was on the cutting edge of call center technology with several products, including its four Rockwell Galaxy ACDs. Eleven years later, the shopping network still runs a very technologically sophisticated call center. And they are still routing calls with those four Galaxies.

"There were two reasons the Rockwell Galaxy was chosen in 1986," says a representative of the shopping network. "One was the redundant design. It is like two switches in one. The other was that it provides detail reporting, from trunks to agents. No one else even came close back then."

More impressive than how the Rockwell switches have held up over 11 years is what the shopping network has been able to do with them — and the company doesn't even use Rockwell's most recent software release. (The company spokesperson explains that many of the new releases contain IVR functions that the shopping network doesn't need with its own very large IVR system.)

The shopping network tapped into the computer-telephone integration (CTI) capabilities of the Galaxy early, using data lines from the switch to perform screen painting.

The ability to share calls between ACD gates and agent groups have helped increase efficiency and reduce costs over time, says the company spokesperson. "Until intelligent routing, that was the most important thing we have done," he says.

Recently, the shopping network added that intelligent routing to its call center through a feature from its long distance carrier. This feature takes real-time statistics from the ACDs and routes each call coming into the center based on those statistics.

Dealing with call spikes has always been the most challenging part of managing the shopping network's call center. An IVR system handles over half the calls. All calls go the IVR and the Galaxy ACDs are located behind them.

It's a little different from the setup at most call centers (where the ACD gets the calls first, then hands them off to the IVR system), says the company spokesperson, but the large volume of calls the IVR system can and does handle makes it necessary.

How does this industry leader handle calls from its Internet business? Its Internet subsidiary is a separate business from the shopping network. The company spokesperson expects a merging of the Internet and television some day, but for now, the shopping network call center does not handle Internet-generated phone calls.

The company spokesperson reports that the shopping network's corporate team is looking into rebuilding the company's entire infrastructure. He and a colleague may rebuild the company's call center solutions as part of that infrastructure change.

They plan to have the call center decisions made by mid-1998, with the entire company project completed before the next millenium. the company spokesperson says that new intelligent routing functions, such as skills-based routing, have put new ACDs on the shopping network's shopping list, but for now there is no hurry.

"The good news is they are not broken," says the company spokesperson, "so we don't have to rush."

CALL CENTER MANAGEMENT

It's difficult to talk about this catalog without getting a little starstruck. It is, after all, a part of a group that is owned by a major movie star.

If the catalog's customers are star-struck though, it is probably by excellent service. The catalog offers home furnishings, clothing and jewelry that are handcrafted by skilled artisans throughout the United States. Callers are handled carefully, with minimum gimmickry.

One of the tools the center uses to handle inbound calls is a 38-station Applied Voice Technologies ACD. The system can integrate with a PBX, notes the call center manager, but the catalog chooses to use it as a stand-alone system.

The call center manager finds the color monitor that displays call statistics very helpful. With the display, thresholds can be set and when they are exceeded, the colors on the monitor change. This alerts a secondary group — not assigned to telephone work — to jump in and help answer calls. "It's a user friendly system," he says.

Phone management software tells him the optimum number of agents and trunks. Something he already figured out by trial and error, he admits.

The display telephone sets shows the agents how many calls are in queue and the average hold time. This information gives the agents the info they need to better manage their time. The center will soon move to a new facility and the call center manager is considering adding a readerboard at the new location to provide the agents with even more information.

About 32 agents work at the catalog's call center from 7 AM to 10 PM. After hours a service bureau handles customer calls. Using the service bureau is an effective way to provide 24-hour-a-day, seven-day-a-week coverage and to provide disaster recovery.

When things get busy in the fall and winter, there are 50 to 60 agents working in several shifts.

The call center manager says the catalog call center is "still fairly small." But the statistics show these diligent agents are getting the job done.

The center handles about 25,000 calls per month, reports the call center manager. The average talk time is three minutes. The average speed of answer is 90% to 95% within 20 seconds.

The call center handles catalog requests and customer service calls in addition to orders. These non-order calls represent 10% of the calls received by the center.

After speaking with the call center manager it is easy to imagine the catalog call center as a place where the calls are as well-crafted as the products they sell.

CALL CENTER MANAGEMENT

When a major retailer began its women's clothing catalog a year and a half ago, it had about 40 agents in its call center and call center managers made up agent schedules by hand.

Today the retailer sends out eight to nine catalogs a year. After each catalog drop the call center receives 10,000 calls per day. Over 100 agents handle those calls.

One of the call center managers laughs at the prospect of scheduling all those agents manually, like they did in the old days. The call center has been using a call center management software system from TCS for the last few months. "There is no conceivable way we could schedule this many agents manually," she says.

The retailer purchased the TCS system in September and have been using it since April. One of the call center managers was impressed by the amount of training she received in using the system. There were three training sessions. First, TCS trainers came to the call center for a few days. A few weeks later, the call center manager went to TCS for a few days. A few weeks after that, the TCS trainers were back in the call center.

The call center manager said she didn't understand the full power of the system until the call center had gathered enough historical data into the system and her training was complete. "They give you homework," she says of her training. "And you had better do it, or you won't understand what's going on."

In addition to the scheduling function, the call center manager makes good use of the exception reports. These reports help the call center managers — and their valuable team managers — track absent and late agents, and other deviations from the schedule. One favorite report is a very visual representation of what the agents are doing. Generated during busy periods, it can be handed to team managers who then can see what needs to be done.

The call center managers says she can tell the new catalog is successful by the telephone calls and the way the customers respond to it. But there may be another measure of the catalog's success: "We're growing quickly, and having a good time doing it."

CALL CENTER MANAGEMENT

Ten years ago a PC products catalog started out with just a handful of people in its headquarters office. Now they are a computer catalog powerhouse, handling up to 45,000 calls per day with up to 400 agents on the phones at any one time.

The company's three core businesses now send out catalogs in ten languages to 13 different countries.

For the last seven years Agent Window from Telecorp Products has helped with rapidly-growing company keep on top of its ACD statistics and its agents.

When the company outgrew its Rolm switch, says its Director of Worldwide Technical Services, it didn't outgrow its need for detailed reporting. The Nortel Meridian provided the switching power, but the company felt the reporting package cost too much.

Although the company is a Macintosh sales leader, seven years ago Macs just weren't feasible in a call center. Agent Window filled the bill for detailed, real-time reporting on a DOS platform.

It's still filling the bill today. The director isn't even sure how many of the Agent Window systems are running in the company's five call centers. "When I need another one, I just call," he says with a laugh.

The catalog company uses Agent Window to improve customer service in several ways. First, the call center has Agent Coordinators who are responsible for 30 to 90 agents. When calls exceed a five minute talk time threshold, the agent's name starts flashing on the Agent Coordinator's screen.

(The average talk time for the company is three-and-a-half minutes. Coordinators are alerted when the talk time exceeds four minutes, but the action doesn't start five minute mark passes.)

At five minutes, the Agent Coordinator checks to see what the problem is. Often, reports the director, the agent has merely encountered a very detailed question or a particularly long order.

"If it seems to be OK," he says, "they will bounce out, but continue to check as needed."

After seven minutes, a supervisor is alerted. If a problem is found, such as an agent not being able to handle a tricky question, the coordinator decides how to best solve it. Sometimes that means the supervisor is brought in for assistance.

The catalog company lets agents manage themselves through readerboard displays of the real-time ACD information provided by Agent Window. the director would like to expand this capability so agents at remote sites can see what the situation is at the host site.

The director thinks Telecorp Product's QWatch could solve this problem for him. In fact, he likes the fact that QWatch puts ACD stats on agents' screens so they don't have to look up at a readerboard.

He hasn't implemented the product yet because his agents are using a mix of PCs, Macs and terminals off host computers and it would be difficult to get them all the QWatch info.

Agent Windows has been a helpful tool for the catalog company. It is clear the director is in the spirit of the product's color-coded, at-a-glance supervisor screens. Describing a good day in the call centers he says, "Hopefully, all you see is green."

ELECTRONIC CATALOGS

An electronics catalog wanted to save money the postage and printing of 10,000 copies of a 320-page electronic equipment cabinet catalog. Postage and printing was $6 for each catalog.

It used diskette-based electronic catalogs and ordering system software from Eon Corporation. The catalog and order software runs on any IBM-compatible PC with DOS 3.0 or higher. It presents up to 120 lines of text and 50 images (color or black-and-white) for each product. Orders are sent from the customer to the company by e-mail.

Using the system, the electronics firm now sends 6,000 diskettes for less than $1.20 each. The catalog is updated by modem instead of through additional mailings. Electronic ordering means no data entry is needed. The system lists active customers, which are given special attention by sales reps.

ENTERTAINMENT
ACD

A ticket sales agency wanted to better serve callers with a selection of sophisticated features, including recorded messages, that could keep pace with the variety of events they take telephone orders for.

Using a small ACD system at each of its two locations in Denver let the two call centers function as one. The system's Extended Messaging option lets the ticket agency give callers info on schedules and ticket availability without sending them to an agent — or using additional call processing equipment.

VOICE MAIL

Staffing a call center is tough. When you rely on volunteers, as one theater in California does, complete coverage is nearly impossible.

But callers to the theater get satisfaction, even when no one is around to answer the telephone. The theater uses a voice mail service from Pacific Bell Information Services that lets callers reserve tickets, inquire about auditions and conduct general business, no matter when they call.

PACKAGED GOODS
ACD

Change can be a good thing, but once the call center at a major seed company, found it very difficult to change. "The telecommunications equipment we had was very inflexible. We had to have the same configuration three years later because it just couldn't change," says a company representative. The company found flexibility with an ACD from Conversational Voice Technologies This system has been able to keep up with the company's growing call center.

The flexibility also has another dimension. "Now making changes has been brought down to the user level. Before we had to call in technicians to do their voodoo magic before a change could be made."

One of the biggest ways the company's call center is growing is by expanding their hours. Theirs is a seasonal business with a busy season that stretches from January to May. March is particularly intense.

During busy season, the order department adds a weekend shift and stays open on weeknights until as late as 8 PM. During these peak months between 60 and 80 agents are on the job. A company representative says the company also hopes to open customer service on Saturday so the order takers don't get tied up with customer service calls.

In the off-peak season the call center hours are cut back. Callers are directed to an 800 number where their catalog requests are recorded by a automated attendant and voice mail system.

The call center manager's busy schedule doesn't leave much time for philosophizing, but she has come to believe one call center truism. "We are working in a world where everyone wants instant gratification, and you had better be on the other end of the line when they call," she says.

The call center is divided into three business segments: home gardeners, commercial gardeners and retail. These segments are further divided into order taking and customer service. Customer service agents resolve customer problems, make adjustments to orders already placed and answer customer questions.

All of the company's agents need a lot of specialized gardening information to answer callers' questions. This makes agent training particularly difficult. "I can't say it ever stops," says the call center manager. "We have to train every day."

Some of this training takes the form of group leaders and supervisors monitoring and coaching the agents, but there is also a lot of formal training. The company has a big telephone training session before the busy season, and the training sessions proceed, each more in depth, from there.

The most difficult task is getting temporary customer service people up to speed in a single season. In addition to the training sessions, the company is developing a database of product and horticultural information to help give the reps the information they need to assist customers.

PUBLISHING
IVR

When a small publisher of telecom books wanted to improve customer service by increasing hours of operation without the need to staff with live agents 24-hours a day, it used an IVR system from Wygant Scientific.

Using the system, after call center hours, calls are automatically transferred to an interactive voice response system that takes the caller's name, country, telephone number and books they are interested in ordering. These messages are delivered to the call center manager's voice mailbox and are transcribed each morning. In the future the system may also give callers the option of leaving a credit card number and shipping address to complete the transaction on the first call. The call center has four agents.

SERVICE BUREAU
ACD

A fulfillment service bureau that responds to literature and material requests for their clients, which include major auto makers, a cruise line and an electronics firm has 12 agents working Monday through Friday. Their ACD is equipped for up to 16 agents and 24 digital trunks.

They switched to the ACD from using two extensions on their PBX. Two ACD groups are answered by live agents who handle requests and process orders. A third group goes to a voice mail system, which allows callers to leave a name, address and information request without a long wait.

The company benefits from ACD features such as multiple agent groups, exception-oriented supervisor monitoring and sophisticated reporting capabilities in a system that serves centers with four to 32 telephone representatives.

TOUCH SCREEN

An inbound service bureau wanted to make its 200 agents "instant experts" on a bewildering variety of inbound clients, including several catalog companies. It used DMI/TAS touch screen and software by Interactive Response Technologies, a subsidiary of CSC.

Using the system, agents navigate through detailed scripts through the software's icons and a touch screen. Information on about 150 different products is stored in the system. Agents touch their way through the ordering process, or to find out the answer to a question.

SPEECH RECOGNITION

A telemarketing service bureau wanted to automate survey and order-entry applications through voice recognition.

It used an interactive voice response (IVR) system from Wygant Scientific that features eight ports of continuous speech recognition and

16 ports of discrete voice recognition. It handles five different promotional programs simultaneously.

The system uses ANI to capture the caller's phone number, speech recognition to capture zip code and product names or other phrases prompted by the system. Names and addresses are recorded and transcribed by an operator. The other information automatically appears on the operator's screen through a Data-Follow-Me function.

RETAIL
ACD

A major camera manufacturer has 48 customer service agents that handle 28,000 incoming and 4,200 outgoing calls per week. The call center is open Monday through Friday, 7:30 AM to 8 PM Eastern Standard Time.

An digital integrated voice response interface allows instantaneous ACD-to-VRU and VRU-to-ACD call handling integration. The company uses it so field technicians can leave messages with a voice processing system. Those messages are transferred to live ACD agents during slow periods.

In the future the integration will allow voice processing system-based meter readings, a voice processing system-based ordering application and an after-hours technician dispatch system.

The average customer hold time was reduced to less than 10 seconds. Other benefits included extending the life of the company's PBX by off-loading call volume to the ACD and voice processing system, and enhanced call information.

TRAVEL
ACD

A travel agency that sells more than 125,000 airline tickets, cruises and tours each year uses an ACD to handle calls effectively.

Calls are transferred to the agent who has been without a call the longest. Call stats tell managers how long calls wait in queue, how many calls each agent handles and how calls are distributed by group.

The travel agency benefits because daily line usage information prevents busy signals and paying for extra telephone lines.

ACD

A major airline carrier uses an Aspect CallCenter with its Network InterQueue feature.

Three airline call centers are linked by a sophisticated private voice network. Network InterQueue allows a call to be simultaneously queued for agents in all three locations. The longest call waiting goes to the next available agent.

What's more, queue and caller information also travels seamlessly between sites. For example, a travel agent's ID number entered in a call to one center, will be transferred, along with the call to another. Call and screen synchronization, and call and screen transfer also work between sites.

The airline has gained a consistently high level of customer service that is not reduced by rapid changes in call volume, such as a TV advertising campaign. The three sites seem like a single center to callers.

ACD

A major hotel chain uses an ACD chock full of features including personal announcements, intelligent delay announcements, eight music sources, ACD mail and automated attendant. Agents at 300 positions handle 35,000 calls a day. The company adds a new hotel every seven to ten days.

While several different toll-free numbers come into a single Rockwell Spectrum switch, the call handling remains distinct, with different music, jingles and promotional announcements for each of the hotel's chains.

Agents can record three six-second greeting in their own voices, which they use to answer calls. These greeting are always fresh, accurate and give agents a moment to collect their thoughts before the call. They can also nip call holding time by as much as four seconds.

Since service changes to the ACD can be made while the system is on line, and because so many functions are now contained in one "box," downtime is practically eliminated.

The hotel company gains an ACD that can keep pace — both in size and sophistication — with a rapidly growing call center.

ACD

A Far Eastern betting club handles up to 35,000 calls in 15 minutes using two ACD systems from Intecom. It has 1,810 agents taking bets over the phone.

Intecom's CallWise ACD function not only lets the club answer a huge amount of calls, but in combination with fiber optics, it allows call handling to take place throughout their dispersed facilities.

The agents take bets in English, Cantonese and Mandarin. A customized digital station set was created for their use. It features a large LCD (liquid crystal display), two built-in headset amplifiers and jacks, and a recorder amplifier.

In the future, an IVR system will collect callers' account numbers while they wait for an agent. The club expects this to cut holding time by 10 seconds and increase the number of calls they can take in 15 minutes to 42,000.

The new system lets the club handle off-course betting for one of two horse racing tracks they own. About 20% of the club's revenue is from these phone calls.

IVR

One of the great perks of working for an airline is you get to fly free or cheap. One of the problems airlines have with this perk is that handling the details of these "non-revenue" bookings costs them money.

One airline solved this problem with an IVR system from Periphonics. Before they developed this system, says their IVR manager, "these calls went into our call centers where they took up inbound lines and the time of our highly skilled agents."

Three Periphonics units with a total of 312 ports handle the "non-revenue" application and two other applications. The three applications are reached through toll-free numbers and dialed number identification service (DNIS) routes calls to the correct application.

When employees call in they enter their employee number, carrier code (employees of 15 express carriers can also fly with the company), and social security number. Using menus, they can check which flights are available for employee travel, list themselves or eligible family members for travel on that flight, verify listings or cancel them.

The employee number is tagged to a database of some 46,000 employee names and the names of employee family members who are eligible to fly at a discount. When a listing is made, a text-to-speech subsystem of the IVR system reads a menu of eligible fliers associated with that employee number. The caller selects the appropriate name from the menu.

Later, if the caller wants to verify a flight listing, the text-to-speech system is able to announce the name of the listed flier.

The airline uses a similar application, running off the same three Periphonics IVR systems, to allow pilots reserve jump seats for flights.

The airline's main use of the IVR system is a flight information system. This popular IVR application accounts for over 17% of their call centers' call volume.

All three IVR applications have been very successful, says the IVR manager. The airline reports a 100% service level answering on the first ring for their IVR applications.

The applications have been so beneficial, reports the IVR manager, that the airline plans to roll out four more IVR applications and another four next year with the goal of having 40% of their calls handled by an IVR system.

CALL CENTER MANAGEMENT SOFTWARE

"I don't know what life was like before the Analyzer, but it must have been the dark ages," says the manager of a Midwestern travel reservation center.

The reservation center is part of the in-house travel agency for one of the nation's largest conglomerates. This office handles the travel plans for most of the conglomerate's automotive locations, and a few other locations that fall within their geographical area.

The company has another in-house travel agency at in the southwest.

The travel agency uses Chadbourn Marcath's CC Analyzer-VS, a package that provides management reporting and an historical database for call centers with less than 30 agents. The agency has 10 agents.

The manager says without the Analyzer he would have to pace the call center floor to try to keep tabs on everyone. Not only do the management reports tell him what he needs to know about his agents, it also makes reporting to management easier.

The manager must submit an operations report every Monday for the previous week. Without the Analyzer, he says, he would only have half the information he needed for the report. As it is, he runs key reports on Monday to give him information about the number of calls, the number of calls answered, the number of abandoned calls and other information from the week before.

He is quite pleased now that these reports have been tweaked to fit his needs. For example, he now has the Success Factors report giving statistics in 30 minute increments in stead of by the hour. He says it gives him a better handle on what's going on.

He also relies on the service level report. He runs off a copy of the report as he speaks. "The company has a commitment to answer 80% of the calls within 20 seconds. I need to know what percent is answered in 20 seconds."

Finally, he says the fact that the Analyzer is Windows-based makes it ideal for the travel business. An agent uses his or her reservation software constantly, he explains, and rather than having two monitors on his desk, the Windows-based Analyzer lets him toggle back and forth between the reservation software and call center information.

CALL CENTER MANAGEMENT SOFTWARE

The reservations center for a major hotel chain has been using a call center management system from TCS for nine years. They know the system so well that last year their analysis of their yearly call statistics lead to a budget prediction that was within 6,000 calls for the entire year.

Not bad for a call center that receives 62,000 calls a week.

The reservations center handles calls for the hotel's franchisees. At last count they were working with 75 hotels, says the center's Staffing Administrator. A total of 200 agents keep the center running 24 hours a day, seven days a week. There is one central reservation center in the United States.

The center handles more than just reservations calls, says the staffing administrator. The center handles the chain's frequent customer awards program and other types of calls.

The hotel chain has used the TCS system for so long, says the administrator, "we've probably used every version of the system." The center has always had TCS' Scheduler, Forecaster and Tracking modules, she says, and they recently added their Real-Time Adherence package.

Real-Time Adherence integrates with the center's Rockwell Spectrum ACD. The package lets the center compare the number of agents scheduled to the number of agents actually logged on to the ACD. An in-depth look shows where the variances are.

The administrator says a discussion of the TCS system can not end with the technology alone. "I rant and rave about TCS' support. It's very important. They don't leave you high and dry."

She is also enthusiastic about TCS' annual users conference. She says the company puts a lot of effort into keeping the costs of attending the conference down, a fact that she and her company appreciate. The conference is a great place to meet with other users and share solutions to common problems. A wish list session means TCS always knows what she would like to see in the system.

The administrator says she couldn't think of another thing she would want from her TCS system that she doesn't already have. It's not because the system was always perfect, or because she lacks imagination. TCS has been very responsive, she says, in keeping up with suggestions from the hotel and other users.

SPEECH RECOGNITION

A leading US travel agency wanted to bring productivity gains similar to those experienced by airlines, hotels and car rental companies to its own business. It wanted to supplement agent-assisted reservations, especially on low-margin air routes.

It created an interactive voice response system that uses a multi-thousand word vocabulary and has speaker-independent continuous speech recognition from BBN Hark.

Using the system, callers dial a toll-free number and speak their identification number. Prompts tell travelers when to say where they are leaving from and where they are traveling to. More prompts ask for date and time of travel.

Callers choose from a list of flights and fares. The reservations are automatically confirmed and the tickets are sent through an office of the nationwide travel agency.

12

SALES

BANKING
CUSTOMER SERVICE SOFTWARE

A bank in the Pacific northwest wanted to increase service, sales support and increase customer satisfaction. It hoped to do this by offering a complete financial service assistance gateway to their clients.

It used a customer contact and sales automation system from Brock Control Systems.

Using software with customer tracking, sales opportunity and account management functions greatly improved service, sales support and customer satisfaction. Lost marketing opportunities are a thing of the past. With the Brock system relationship marketing is the order of the day.

CABLE
DIALING

In the cable television industry, instant gratification is usually associated only with pay-per-view. But customers of a Florida cable company can get instant gratification on upgrades to premium services too, thanks to a form of computer-telephone integration.

The cable company uses a Melita PhoneFrame predictive dialer to call their existing customers for upgrades to premium services (like HBO and Cinemax) and to sell pay-per-view broadcasts.

Before purchasing the Melita system, the cable company agents dialed their calls manually after receiving a stack of sales cards for each campaign.

"There were different cards for different campaigns," explains the company's call center manager. "The sales reps had a lot of cards and had to do everything manually."

With the Melita system in place the dialer and a computer system take care of the dialing (screening out busies and no-answers in the process) and assist with the record keeping.

Agents know how many calls they have made and how many people they have contacted before they switch to the next campaign.

Before they used the predictive dialer, the agents averaged 14 presentations an hour using the manual system. Now they make about 25 per hour.

What makes this application sizzle is hot-key access to various software functions right from the agents' terminals. A Melita PC interface forms the link between the dialer and the software.

This link gives agents access to the most up-to-date information when they reach a customer.

For example, says the manager, without instant access an agent might tell a customer that adding Showtime will only cost $5.60, because his or her records show the customer subscribes to HBO.

If the customer dropped HBO after the records were downloaded, but before the call was made, the information is inaccurate. The customer will be billed the higher, Showtime-only rate, which will probably result in a call to customer service about a "billing problem."

With up to date information, the agent sees the customer doesn't have HBO and quotes the correct price.

But hot-key access to the software also means customer orders can be entered right away. Before the integration customers would have to wait a day to receive their new service because their order had to be processed later on another system.

Now agents can access that system and customers who order Showtime (for example) get it immediately.

This integration means the cable company's predictive dialer doesn't just deliver more contacts per hour. It also delivers superior customer service through accurate, on-line information and instant access to important functions like order logging.

DIALING

"Right now cable doesn't have any REAL competition, but as the different transmission technologies mature, they will," says a representative from a popular cable channel.

Cable, meet the competition. This entertainment company not only sells its programming through local cable companies, but it also goes direct — to users of satellite dishes.

The call center is one of the cable station's points of contact with satellite dish owners. Right now satellite dishes don't really compete with cable. The dishes are too large and too expensive. As the dishes get smaller and less expensive, though, satellite dish service will become more competitive with cable.

The cable station uses a predictive dialer from TeleDirect to call customers.

The station spokesperson would like to be able to reach prospects with the predictive dialer as local cable companies do. But the cable station's calls are limited to existing customers because calling lists are scarce and expensive, the station spokesperson explains.

The cable station uses the dialer for two functions, annual renewals and collections. Collections are common to most businesses, but annual renewals are something local cable companies don't have to deal with.

Contracts for satellite services, such as the cable station, are not open ended as are subscriptions to the same services through a local cable company. The cable station sells annual contracts and must renew each of those contracts each year.

Usually the cable station sends out a mailing, then calls. On the renewal call they may also try to upgrade customers, to add another service like Cinemax, for example. Occasionally they will run upselling campaigns.

Before 1993, when the cable station started using a predictive dialer, all of these calls had to be made manually.

"It was a tedious process," says the station spokesperson. "The sales associates had to highlight different calls in different colors on their sheets. They had to keep track of when to call back. Productivity wasn't as high because the sales associates could not focus on making the sale."

Now that calls are made automatically with busies and no-answers eliminated, call-backs can be scheduled without sales associates watching the clock, and records of calls and contacts made are tracked by the system, sales associates can concentrate on communicating with the customer.

The results have been phenomenal. The per-hour renewal rate doubled since the cable station started using the system, the station spokesperson reports.

The call center also handles inbound calls using a Nortel Meridian Max telephone system. With help from a company subsidiary, the two systems are now integrated.

The integration means sales associates can access the dialer using a button on their Nortel telephones. Having just one station set on the desk is very helpful, says the station spokesperson.

Before the systems were integrated the call center was cluttered with additional equipment. The new arrangement gives everyone more room. "Its a very clean operation," he says. "Switching between inbound and outbound requires just one button."

As a provider of satellite TV programming, the cable station is on the cutting edge in more ways than one. With improving technology, satellite dishes are expected to be an increasingly popular way to receive TV programming. As this happens, more of the industry may be looking to methods the cable station uses to serve their customers in the future.

DIALING

A cable company with 350,000 subscribers wanted to save time, increase productivity and control costs. It used a predictive dialing system from EIS.

Using the system, response rate jumped from 12% to 25%. Cost per call dropped. Agents no longer shuffle computer printouts and dial numbers manually. The automated system dials numbers and tracks calls.

HEALTH
SALES SOFTWARE

A health products company with 15 field sales people and two regional sales managers wanted to provide a better flow of information between the sales force and the administrative personnel. The wide range of products and services offered by the company make this a particularly complex task.

To do this, it used MarketForce, a sales and marketing automation software system from Software of the Future.

Using the system, salespeople download client information into the company's distributed database system in the head office overnight. The following morning managers can act on the new information. The two-way flow of information includes reports, Lotus files and spreadsheets to create a complete fact file remotely.

The company feels moving information is now faster and more efficient. Management is confident of reaching the right people, and the sales force has the support necessary for the day-to-day management of their customers.

DIALING

A Medicare risk contractor provides health care by contracting with independent physicians and medical groups, as well as owning and operating some of their own centers.

The contractor wanted to use direct mail and telemarketing to increase sales to new customers and improve the productivity of their field sales force. First done manually, these applications were automated in 1994 with Edge customer management software from IMA.

The Edge software helps the center track campaign response rates by tallying calls to each of the center's toll-free numbers. Potential customers who don't reply are called using a predictive dialer. Each of the company's 104 agents use the Edge guide to verify information, ask qualifying questions and set up an appointment with a field salesperson.

Overall productivity has increased by about 60%. Paperwork — including call lists, appointment calendars, prospect comments and statistics — has been virtually eliminated. The system has even automated the sending of personalized letters and literature. Call tracking capabilities lets risk management company measure response rates on each marketing campaign. Reports give them info on each rep and calling group.

IVR

A Boston area HMO wanted to offer preferential memberships rates with the a local health club to encourage fitness, and therefore better health, in their customers. They wanted to be able to handle the some 5,000 calls per day generated by the program, and make it easy for members to participate.

To do this it used an interactive voice response system (IVR) developed by HTI Voice Solutions.

Using the system, callers can zero in on the nearest health club by location. With a member number identification, the registration is recorded and processed. Information is sent to the selected healthclub site, and to a mailhouse, which sends that member a postcard confirmation.

The health club Enrollment Line is available to health plan members 24 hours a day, seven days a week.

HIGH TECH
SALES SOFTWARE

This company is one of the world's biggest PC distributors and systems integrators. They are a Fortune 500 company with annual sales of over $2 billion.

Until recently, they didn't use a contact management software system. Not surprisingly, when they decided to automate this aspect of their business they chose a comprehensive customer service management system that met their sophisticated needs.

The call center at the Arizona-based company has just over 100 agents, who are split into two groups.

One group is customer service. The agents in this group are called "account managers" and they handle relations with people who resell the company's products. The account managers receive mostly inbound calls, but they also make a few outbound calls.

A business analyst in the company's IS organization estimates the breakdown of the calls the account managers receive is 80% inbound and 20% outbound.

Account managers are customer champions who handle all aspects of the company's relationship with the reseller. They are the people the resellers can turn to for information about orders and back orders, and they help solve any problems the resellers have.

But the relationship doesn't end there. Account managers may make sales calls to the resellers or they may discuss a new product or promotion with a reseller when he or she calls in for another reason.

The other call center group is the commission sales force. They are dedicated to the company's value-added resellers. The commission sales force is more aggressive about making outbound calls to their client base. Still, says the analyst, about 60% of their calls are outbound and 40% are inbound.

The typical call in the company's call center is an inbound call from a reseller. Callers always deal with a member of their account management team, so when the call comes in it is queued for the next available team member.

The analyst says typical transactions include checking on open orders and back orders, following up with outstanding issues and placing new orders. While the caller is on the line the agent will check to on little things like telephone number or address changes and other administrative details.

The company's primary computing system as an Amdahl mainframe running MVS and CICS 3.3. Their telephone system is an ISDN-compatible Lucent G3r.

The company had never used a contact management software system before they installed an Early, Cloud customer service management software system nearly two years ago.

In addition to working on and with the computer mainframe and the telephone system, the Early, Cloud software runs the company's Real Fax processor. The company sends out a magazine to clients. The magazine includes a very short product description and a document number for more information.

Clients can call into a fax system, enter the document number and a PIN and have detailed information on the product, including pricing, faxed back to them. The PIN number places some limits on who can get pricing information.

Last but not least, the system is integrated, through the mainframe, with e-mail. The Early, Cloud software can automatically send customers e-mail messages.

"Before we installed Early, Cloud our customer information databases weren't well integrated," says the analyst.

Because the company agents work in teams, and because many the company departments are all diligently collecting information on the same customers there can be too much information in the system.

The problem, says the analyst, was that it was possible to have 10 versions of the same basic customer information such as name and address, simply because the same information was entered differently by different departments.

The company needed to keep the information consistent between departments and needed to get information quickly to everyone who was working with a particular customer. They realized that a more effective distribution of customer information would mean better service.

The Early, Cloud customer service software system selected by the company after a detailed and careful search has several major functions. The core of the system lets the company develop applications and provides workflow management.

"Workflow" has been described as an automatic, electronic "to do" list, which makes sure computer files are delivered to the right person's computer at the right time, telephone calls are made when scheduled (or necessary) and that actions are completed and recorded by specific people at specific times.

The system also integrates data, letting the company combine those competing database into a single, comprehensive one. Outbound calls, in either predictive dialing or power dialing mode, can be generated by the software system.

There are also functions that generate forms or letters and an ISDN function for advanced features such as screen pops. The company doesn't necessarily use all of these functions.

They have taken advantage of Early, Cloud's all-encompassing database to handle promotional mailings and other customer communications. All mailings are generated through the Early, Cloud system, says the analyst. Customer information is used to select customers for a campaign and to customize the mailings.

"We did an extensive search to match our needs to what was out there," says the analyst. "One of the reasons for selecting Early, Cloud was that it was easily integrated with our other distribution and logistics [software] systems."

The company has a very high volume of transactions, so they must work on a mainframe. "Because in the PC industry everything turns on a dime, the software system had to be very flexible," she says.

Early, Cloud was the solution of choice because it could handle the company's transaction volume, while remaining flexible.

The analyst makes it clear that the first step in implementing the software system came long before they bought or even searched for a product. the company made a careful analysis of how their call center does business.

Agents were asked for their most common tasks, and those tasks were streamlined. "You must keep in mind your goal," says the analyst. "The first objective is to make the sale. Design the system so your agent can concentrate on that, and not on trying to navigate the software."

The analyst warns that a system as flexible as Early, Cloud's can be treacherous for a company that hasn't studied their process flows. It is easy to get carried away, she says, but you must keep in mind that the system must be easy to use for your agents.

The company took a number of approaches to this end. They standardized just about everything about their many Early, Cloud-based applications. No matter what screen or what script the agent is in, the PF keys perform the same functions and the colors are standard.

For example, no matter where the agent is in the system, editable fields appear in green on the screen. Protected fields are in blue. "This really streamlines training," notes the analyst.

The company also kept in mind that the eyes move vertically when they are scanning quickly. They designed screens so the most important things were arranged vertically, sure to be noticed even if the call is a quick one.

"The biggest thing is to build your system with as much intelligence behind the scenes as you can," says the analyst. She says the ability to do this is one of the biggest benefits of the Early, Cloud system.

This can be a daunting task, but, she adds, "If you really understand your call flow and your back-end it is possible."

The care and attention the company put into designing this system has paid off handsomely. The system is finely tuned to the way the company does business.

For example, when the company's call center system dials an outbound call for a sales campaign, the call is queued to the account manager who works with that customer — keeping with the company's policy of having the same account manager handle all contacts with a customer.

There are also more tangible benefits. The company has saved $250,000 by using workflow for their vendor authorization process. The system makes sure phone calls get made and information is passed on.

The company also uses workflow to manage activation and deactivation, and their quality assurance process. In quality assurance workflow automates the escalation of the customer issue, and serves as an issue tracking system. Because the system is company-wide, anyone one can take care of the problem.

The analyst is working on an ISDN project right now that will bring screen pops to the company's call center in six to nine months. It will take advantage of Early, Cloud's ISDN function. A caller's Automatic Number Identification (ANI) information will trigger a database look-up, and the caller's customer information will appear on the agent's computer screen as the call is answered.

"We're definitely on a migration path to ISDN," says the analyst. With the attention to technological detail the company has already shown, and their powerful new customer service software, this will mean even better customer service for callers to the company's call center.

PULLING IT TOGETHER

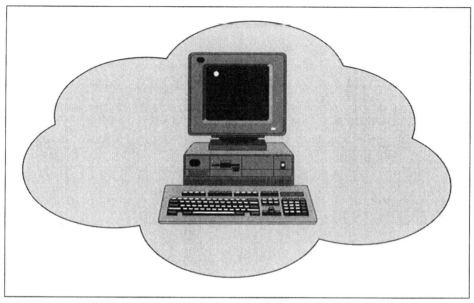

The company had 15 different customer databases before they implemented the Early, Cloud system. By integrating the Early, Cloud system with their databases, they created a single, centralized database used by all the company's telephone agents. This integration is possible even when client/server LAN systems are used in addition to the mainframe system which serves as host to the Early, Cloud software.

CALL ACCOUNTING

In some call centers, agents don't know how well they are doing on the job until their first review, six weeks after they are hired. At one PC training company, headquartered in Minneapolis, agents and supervisors know how things are going right away, thanks to a call accounting system from IntegraTrak.

The company offers PC software training to corporations and government offices. Their 65 telephone salespeople make calls from an Executone PBX.

Because of the company's innovative use of the IntegraTrak system, supervisors have a profile of call statistics for successful salespeople. By looking at the statistics on reports from the call accounting package, the supervisors can tell very quickly if a salesperson will likely find success at the company.

Of course, the supervisors don't expect rookies to act like veterans from the start. The profile shows how call statistics change with experience. For example, they have found that new employees start off making many short calls. Then, as they build a customer base they make fewer, but longer calls.

The call accounting system is used to find out the number of outbound dials, the number of calls by each staff member and the durations of calls. This information helps them manage the call center and develop staff skills.

PUBLISHING
ACD

Three newspapers owned by a nationwide media conglomerate have turned to Rockwell's Spectrum automatic call distributor to route calls for departments that have a high volume of incoming calls, including circulation and classified advertising.

A system manager at each newspaper takes care of the day-to-day operation of the ACDs, but one woman at the conglomerate's headquarters oversaw the selection of the ACDs at these three newspapers.

Before purchasing their Spectrum systems, each of the three newspapers was using the ACD features of an older PBX to route their calls. (At one newspaper, two PBXs were being used.)

The conglomerate's telecom director says she had three criteria while shopping for an ACD for the first newspaper.

- A platform that the manufacturer would continue to enhance to meet the newspaper's future needs.

- A cost-effective investment.

- A manufacturer that would provide local support and service.

The cutting-edge features of the Spectrum, such as skills-based routing, impressed the telecom director but not as much as the fact that the company could purchase such a feature-rich system for a reasonable price. "We could buy more for our money going with the Spectrum," she says.

While the three criteria were important, the telecom director also considered the special needs of each newspaper. In addition, she considered the availability of local support. "In my opinion, service and support is particularly critical. If there was an area in the country where we couldn't get local service and support, it would effect my decision."

Interestingly, the telecom director does not see different ACD needs among the various media outlets owned by the conglomerate. The other medium with intense inbound-call handling needs is cable television, she says, and based on the all-around functionality of the Rockwell Spectrum, she could see the Spectrum serving one of the company's cable TV outlets just as well as it serves their newspaper outlets.

ACD

A newspaper agency in the West serves two large newspapers with a total of 43 agents, with 24 agents working at any one time. The agents handle about 2,000 calls per day.

The newspaper agency wanted to organize and prioritize calls coming in to a busy classified advertising department. It used a small ACD system.

Deadlines are everything in the newspaper biz — even when it comes to classified ads. The classified ad department's busy days are Thursday and Friday.

The ACD's Time Tables feature plays a message at critical times related to ad deadlines. For example, on Thursday afternoon a caller hears a recorded message saying the deadline for the ad he or she wanted to place was that morning. The caller hangs up before reaching an agent — which is good, because the agent is really busy doing other things related to the classified deadline.

Similarly, the agents' Digital Display Sets tell agents exactly when the call they are answering came in to the system, so callers can't claim that they really called in before the deadline and were on hold until after it.

CUSTOMER INTERACTION SOFTWARE

A division of the US's largest newspaper publishing company wanted to improve service to their existing advertisers, qualify new prospects and increase overall sales. Goals any company can relate to.

It accomplished these goals with a sales and customer service software system that it used to build an "opportunities database" with active accounts and prospects. When the newspaper purchases a new list from outside sources, this information is added to the database and the prospects are qualified.

Prospects and customers are tracked by SIC code and geographic location to identify sales opportunities for special advertising sections.

The newspaper's Telesales group uses software to track and manage smaller adverting opportunities. The inside sales group has streamlined their follow-up process. The whole sales department uses the system to maintain its existing customer base.

RETAIL
SALES AUTOMATION SOFTWARE

A business forms company wanted to increase customer contact and provide more efficient service to service stations, body shops, farm implement dealers and automobile-related businesses in rural areas.

The forms company achieved this through an aggressive telemarketing campaign that used the massive database and customer follow-up capabilities of a sales automation software system.

The application gave them a detailed view of customer inventory levels and order history where they had very little information before.

The results are astounding. The company increased their business in this market by 31% in just one year. They now have accurate profiles of these rural-based businesses and are able to develop value-added solutions for them.

TRAVEL
ACD

You see an advertisement for a major ski resort in a magazine or on television. You are in the mood for a ski vacation (we'll pretend it's not May), so you call the toll-free number. That number is answered by the resort's reservations call center.

The ski resort's reservations center handles every aspect of your ski vacation, from your lodging in one of about 90 facilities, to your air transportation, ground transportation, travel insurance and, in some cases, your ski school reservations and ski equipment rental, says the resort's operations coordinator.

During peak season (January) 18 reservation agents and a support staff of 15 handle 1,500 call each day. The center is open seven days a week from 7 AM to 7 PM. An automated system gives the latest snow conditions and another system gathers names and addresses for brochure requests.

A standalone ACD helps keep the center humming. This system recently replaced an older ACD from the same manufacturer, reports the owner of the company that sold the system.

The new system has several advantages. First, it has a voice mail system, so the ski conditions line and the brochure request line, which used to run off answering machines, are now handled on the voice mail system.

"They used to have to go over to the machine to collect names and addresses to send brochures," explains the vendor. "Now anyone can do it from their station set."

The new ACD system is integrated to the resort's PBX, a Lucent Definity. This makes it easier for agents to transfer calls that come in to the toll-free number to other departments. "It lets us cut out the calls that are not sales calls," says the operations manager.

For example, if a caller just as questions about a ski school program, but is not making a reservation at that time, the reservation agent can now transfer the call to the ski school.

The most recently added feature is speed dialing. With the old system agents had to manually dial hotels and other lodging facilities when they needed information. Now hitting a speed dial button initiates the call.

ONE CALL DOES IT ALL

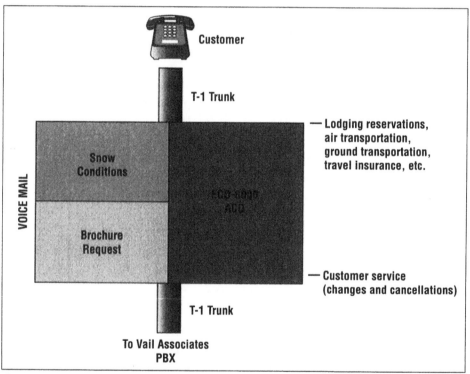

Several toll-free numbers terminate a major ski resort's reservations call centers. By looking at the trunk number displayed on their station set, agents can tell which toll-free number was dialed. Brochure requests are handled by voice mail, is a snow conditions recording. This gives agents more time for sales calls.

The vendor says in the future an automated attendant may let callers route themselves to other departments, like the ski school, during busy periods.

Another change may be the addition of dialed number identification service (DNIS), which would let all available channels on the center's T-1 trunk handle calls for any of their 800 numbers. This would mean fewer busy signals, and agents would still know what type of call was arriving.

DIALING/CONTACT SOFTWARE

One Atlantic City hotel and casino knows how to keep them coming back for more. The casino does this by keeping in touch with their customers through a combination of direct mail and telephone contact. A key component of their customer contact system is its telemarketing software and dialing system from Digisoft.

One promotional program converts new customers into repeat customers. Information is gathered from the 3,000 or so new customers who visit the casino each week, says the Director of Casino Services. That information is entered into the system.

A few weeks later a letter is sent to the new customers with a special offer good when the customer responds by a certain time. A call-back date, several weeks after the letter is sent, is entered into the Digisoft system. About 90 of the customers respond each day. Those who don't are called by agents through phone calls scheduled automatically by the Digisoft system.

The director says the casino has good evidence this system works. Some new customers don't give the casino a phone number on their initial visit. They are not called as part of the program and serve as a control group. Those who do give a telephone number and who are called, visit again at a much higher rate than the control group.

The new customer program is one of many run by the casino. Casino customers regularly receive special offers based on their level of play, their interest and response to certain types of promotions, and their interest in entertainments or games that the casino predicts will be slow at a particular time.

For example, the casino can retrieve a list of dollar slots players from the Digisoft system and call them with a special offer when they predict the dollar slots games will be slow at a particular time.

The director really appreciates the flexibility of Digisoft system's scripting function. Not only does the casino run a variety of promotions that might run a single script page or 40 script pages, but a single promotion (say for a discount on tickets to a show) can have various levels based on the customer's level of play (that is, how much money they spend in the casino).

High level customers may be offered a free room along with the show discount. Another level of customer might receive only the show discount. The Digisoft system pulls up the appropriate script for the customer when the agent accesses information on the customer.

RETAINING NEW CUSTOMERS

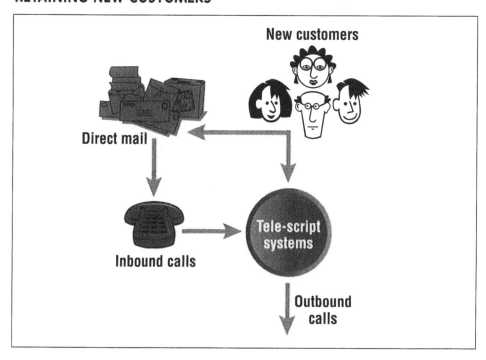

Information about new casino customers is added to the company's database. From there it is used to generate a direct mail campaign, and a call-back campaign that is automated by the software system. When customers respond to the mailing, agents access their info through the Digisoft system. Customers who don't call in response to the mailing get a follow-up call generated by the Digisoft system a short time later.

The director reports the agents dial directly from queue. "As soon as they finish one call, another call is dialed by the system and another script is generated," he says. Using this system has more than doubled the agents' productivity. Previously, agents called up customer records individually from the casino's minicomputer and dialed calls manually.

The system has worked so well for the casino, in fact, that they plan on increasing the number of employees who use it from 22 to 37. The new users will be casino "hosts," who are assigned "players that warrant special attention."

These hosts are experienced, top-notch casino employees who are assigned high level customers as their own accounts. Each host has about 1,000 accounts. The director feels the information in the Digisoft system will help the hosts give their customers the highest level of service.

In addition to the great service Digisoft helps the casino offer customers, it also helps managers keep tabs on their employees. The casino uses the Digisoft system's many reports, including how many calls each agent makes and the results of those calls to gauge employees' performance.

With all the parties and special events sponsored by the casino the fun never stops. And with all the marketing those events require, the uses for Digisoft software never stops either.

13

CUSTOMER SERVICE

BANKING
CALL CENTER MANAGEMENT

Call center managers spend a lot of time worrying about how long customers are kept waiting, but they never think about how long they themselves wait.

Before the call center at bank in Tennessee started using systems from Telecorp Products they usually waited a few days to get their ACD statistics from the telecom department.

Now, for the average corporate telecom user a few days is unbelievably — and possibly unnecessarily — fast, but for a call center it's a long time to be kept on hold.

With ACD Performance software, Agent Window and two readerboards from Telecorp, the bank's call center "can tell at every moment what is going on," says an Assistant VP of the bank's Retail Division. "When we are targeting certain goals, supervisors and staff manage their time better by seeing the information right away."

The bank call center's 27 agents are split into two queues. One queue handles branch support, answering questions from the bank's branches about specific processes and procedures. They have a lot of technical knowledge.

The other queue is called the "Info Line" and is front-ended by a

Syntellect VRU. Live agents handle the tough questions that can't be answered through the automated system. They handle questions about checking and savings accounts, and on several types of loans including equity, installment and commercial loans. This queue also handles inquiries generated by the bank's advertisements.

The two queues handle about 2,000 calls a day.

This environment doesn't need any extra pressure. Instead of lights that flash or sirens that blare when an important service threshold is exceeded, that statistic glows in red on the readerboard for the troubled queue.

The statistics displayed on these readerboards include number of calls in queue, the longest wait and the queue number, says the Assistant VP.

In the near future, the call center will put its VRU behind an ACD to make sure each of its 60 lines is being used effectively. This strategy will also let the department do more number crunching on the reports from the VRU through the Telecorp system.

CALL CENTER MANAGEMENT/MESSAGES ON HOLD

A large bank's mortgage unit wanted to shorten unacceptably long wait times when low interest rates caused a surge in call volume.

The bank used Chadbourn Marcath's Call Center Solutions CC Announcer. CC Announcer announces the current average wait time to callers in queue. Callers can make an informed decision whether to hold or call back, says the company's VP Customer Service.

There has been a significant decrease in abandonments and the flow of calls to the center is more even. The company says this is due, at least in part, to the CC Announcer system and the positive initial reaction it creates for callers.

MESSAGES ON HOLD

A bank in the mountain West wanted to inform, educate and entertain customers while they wait on hold. It used music and messages on hold from The Hold Company in 12 of its locations.

Sample message: "At Our Bank, we want your banking to be convenient. So you'll find that all of our banks offer Saturday hours. Hours do vary from location to location, so check with your local Our Bank. Please continue to hold. A representative will be with you shortly."

Using the system, the bank had positive comments about the service and plans to add it to more locations soon.

IVR

A bank in the great white north wanted to handle sophisticated banking functions such as funds transfer between accounts, interest earned or paid information, and merchant credit card authorizations while spending less on calls to customer service reps.

It used a 32-port IVR system from Brite Voice. Using the system, total call volume increased, but the number of calls handled only by the IVR system increased by 14%. The bank saved money by reducing the number of calls handled by agents by 14%.

IVR

Somehow when you picture paradise you don't picture people in so much of a hurry that they don't have time to go to the bank. Yet in Hawaii that is clearly the case.

A large Hawaiian bank has assets of over $7 billion and 64 branches. The bank's a twenty-four hour a day, seven day a week automated bank by phone service is only one of several delivery channels used, but it is one that has proved popular with customers.

The vice president and manager of the bank's Research and Development Department reports the bank has been offering bank by phone services since 1980. The bank's recent expansion of the IVR line, based on a Syntellect VRU, is part of a processes of continually improving customer service.

"70% of our customers elected to use the voice response unit, even though we have done very little advertising," says the VP. Call volume has increased since the system was expanded. In one branch the number of calls more than doubled. "That's a lot of pent-up demand."

The VP says the new system has improved customer service on several fronts. First, callers receive faster service when they call their local branch. With only so many employees available at a small branch, its natural for callers to be left on hold while in-branch customers and other callers are helped.

The bank's goal is to have the automated system available in every branch so the majority of callers don't have to speak to a live person.

When callers need to leave the automated system and speak to a live representative, they benefit from another level of improved service. The Syntellect system is integrated with Early, Cloud software running on a Tandem computer and a Nortel PBX.

Information gathered by the VRU is ported to the Early, Cloud screen and presented to the representative along with the call so no time is wasted. "It's a neat mousetrap," says the VP.

The third benefit is that with all the time saved by automation, representatives have more time to sell customers on additional bank products. "We believe every call is a potential sale," he says. A fringe benefit is that all of this customer-coddling service and added time for sales is possible with a minimum number of live representatives.

The end result of all this cutting-edge technology for the bank's customers may be the discovery that in paradise, even banking is a pleasure.

CABLE
ACD STATS

A cable company on the Great Lakes wanted to monitor ACD statistics to assure that federal and in-house standards of service are met. They used several ACD packages designed for smaller call centers.

The cable company has been using these ACDs for eight years. They provide information such as the percentage of calls abandoned, the average length of call, number of calls directly answered and much more. The company also uses the ACD to track the number of installations requested and scheduled for their subscriber services.

FINANCIAL SERVICES
ACD

What would you do if a company hired you to take care of the needs of its owners? That is exactly what happens at one of the nation's largest corporate trust group.

As a stock transfer company the trust group handles stock-related inquiries for a variety of corporations. In dealing with stockholders they are dealing with the company's owners.

A key call center technology in this important task is an ACD.

The group has found the vital tools for handling these calls is Dialed Number Identification Service (DNIS), whisper prompts, IVR integration, excellent ACD reports and intelligent, well-trained agents.

All calls are answered first by the ACD. DNIS, used when the client company has its own toll-free number, lets the switch know what to do with the call. Group clients are given the option of having their shareholders sent to an IVR system first.

If they choose this option, the ACD will deliver their shareholders to the IVR system.

A client may also choose to have calls answered directly by a live agent. These agents get a whisper prompt from the ACD with the name of the company the shareholder called. This allows the agent to answer the call appropriately.

An Assistant Vice President for the group says they have found that trying to identify callers though Automatic Number Identification (ANI) is not fruitful. Callers phone from too many different locations to make the application practical.

Teknekron Infoswitch's reporting package is also helpful. It allows the group to forecast call volume and move people around to meet the need. The package also provides additional statistical reporting and a digital display that keeps everyone up to date on the latest stats.

Using this technology and their training, the group's 80 representatives handle 5,000 to 6,000 calls each day. During busy season — which starts in January when people have a lot of tax questions — they may handle 10,000 calls a day.

They do this with the confidence that they have what it takes to make their customers' owners happy.

SPEECH RECOGNITION

One of the largest issuers of Visa and MasterCards in the US wanted to reduce the number of people calling their IVR service system who did nothing at the call prompt — 19% of their callers. The company was aware was aware that 30% of US phones have rotary dials and some disabled customers have trouble pressing telephone keypads.

This credit card company handles 90,000 per day.

The solution was a Lucent Technologies Conversant IVR system running voice recognition software.

The number of callers who do not answer at the prompt was reduced from 19% to 2%. Those customers are saved from having to wait for a customer service associate. A customer service improvement for both caller and company.

GOVERNMENT
ACD

Can offering better customer service be as easy as buying a new telephone switch? It was for one state's Registry of Motor Vehicles.

A new Mitel fiber-distributed PBX helped them cut abandoned calls from 20% to less than one-half of 1%.

Another important part of the package: Mitel ACD Supervision, which helps the Registry track and route calls better.

HEALTH CARE
ACD

A hospital wanted to extend the reach of personalized patient-care beyond the walls of medical center. To do this it used Intecom's advanced multi-media network.

Using the system, the hospital's call center, staffed by nurses and other trained professionals, handles telephone triage, out-patient follow-up, physician referral, emergency patient counseling, discharged patient surveys and keeps citizens informed about community health care programs.

The call center handles more than 150,000 each month. Those calls are 80% inbound and 20% outbound.

The call center is linked through the hospital's computer network to other outreach services, both nationally and locally. The Intecom system provides and integrated platform for voice, data, text, graphic and video services.

PREDICTIVE DIALING

A managed care company wanted to increase prescription compliance in patients on chronic medications by calling them to remind them to have their prescription refilled.

To do this it used Melita's Customer Care predictive dialing system.

Using the system, a sample of the patients who received reminder calls are also called monthly to see if they actually picked up their the medications. In this program, the rate of patients doing so surpassed industry standards by 20%. Of the 64,000 patients reached in a three-month period, almost 41,000 agreed to have their prescriptions refilled. The company anticipates expanding to 96 reps making 120,000 calls per month. Eventually they will reach nearly two million patients using 174 predictive dialing stations.

The call center has 25 customer service representatives and makes 40,000 per month.

HIGH TECH
ACD

A business machine company wanted to replace an overloaded PBX/ACD combination. It also wanted to improve call center information, especially after centralizing all service dispatching to one national location. The company wanted to manage incoming calls better, get help with heavy traffic periods and let field technicians spend less time on the telephone.

It switched to a stand-alone ACD system for the call center.

The new ACD let the company hold on to their PBX for three to five more years. The improved call information led to improved agent work schedules and the average customer hold time was reduced to less than 10 seconds.

An integrated voice mail system from VMX helps field technicians get off the phone faster. The voice mail system takes messages from the technicians, and when things are slow, delivers the message through the ACD to a live agent.

With the voice mail system in place, the company is poised to begin several new applications, including a VRU meter reading collection application, a VRU supply ordering application that will supplement the company's automatic system, and some day, after-hours service.

The company has 48 customer service agents. 25 additional agents are added during the January to March peak period. The center handles 28,000 incoming and 4,200 outgoing calls per week. Calls are answered Monday through Friday from 7:30 AM to 8 PM Eastern Standard Time.

PACKAGED GOODS
ACD

A major breakfast cereal company has an 800 number on each package. Eighteen agents answer customer questions about its different cereals.

The phone calls they receive are varied, says a company spokesperson. They get calls asking where customers can find a favorite product, general comments, recipe requests, opinions about products or promotions, comments about new product introductions and requests for information about promotions.

Helping them handle these calls is an Aspect CallCenter ACD. The older ACD the company used previously didn't have many of the features their Aspect ACD has.

To the company, some of the key features of the CallCenter are:

• The flexibility to let several functional departments use the features to their advantage. That is, features can be customized by department so not every department has to use the same setup.

- Call-waiting monitors alert representatives to callers waiting and long queue times.

- A message auto-callback feature that helps them equalize peak and low periods. The cereal company finds this feature particularly useful.

- Excellent tracking of incoming call data to find staffing problems. Information is available for very short time periods (15 minutes) or much longer ones.

RETAIL
IVR

A major men's clothing manufacturer wanted to avoid losing calls from customers and sales reps who call about orders, but who must hang up before their call is answered by an agent.

It used three Syntellect Call Center Gateway systems and MCI's toll-free service.

The manufacturer's call center answers calls from individuals who purchase its clothing, large retail companies and their own field sales reps. The toll-free service delivers ANI (Automatic Number Identification) information to the Gateway system. An agent then uses that info to return the call.

The Syntellect system routes calls based on the toll-free number called (DNIS) and ANI. With the new system, the manufacturer handles 12 calls per toll-free number. Previously they were limited to four calls per number.

TRAVEL
CALL CENTER MANAGEMENT SOFTWARE

On the strip in Las Vegas there are many spectacular water shows. Inside one of Vegas' largest hotels, a 2,688 room resort, another spectacular show takes place as 16,000 calls are handled each day.

The resort's Telecom Manager says there are eight ACD groups in the hotel running on a Northern Telecom switch, including casino marketing, dinner reservations, room service, the front desk and room reservations.

The room reservations call center has 15 to 20 agents and handles 5,500 to 6,000 calls a day. That's a lot of calls, but in this city where day runs into night, the call center is open a little longer than most — until 1 AM some nights.

The reservations center uses Telecorp Products' ACD Performance Software, Agent Watch and LED readerboards to keep on top of the action.

One thing the telecom manager appreciates about the Telecorp system is the ability to archive historical ACD information. Just working with information from the switch, you can print information every hour, but then it's gone, she says. "With this system we can go back until when the doors opened." This lets the resort do things like compare the number of calls to the same time period in a previous year.

A tape backup means that all that information is not clogging up the hard drive.

The Room Reservation Day Shift Supervisor says the system is a productivity tool, which lets him compare the agents' reports on their conversion rates to the number of calls the system says they took.

The LED readerboard that hangs over one of the doors in the call center tells everyone how many agents are on the phones, how many calls are in queue and the length of the longest call in the queue.

He says during slow periods agents work on other projects, but when the stats on the LED board change from green to yellow to red and the audio warning sounds, they know to hop back on the telephone.

Both the telecom manager and the day shift supervisor note that they can use the call volume statistics from the early part of the week to forecast what the rest of the week will be like. They will customize their staffing assignments based on these statistics.

The day shift supervisor says the Telecorp Products story would not be complete without mention of their quick service. He says their ability to access the system by modem and solve problems over a telephone line is very helpful.

UTILITIES
POWER PROTECTION

A New England power company's telecommunications network is substantial. It supports 15 offices and three call centers. This network has to work 100% of the time. Imagine calling the power company to report an outage and getting no answer.

The power company uses two types of products from Gordon Kapes to make sure its call centers are always available to take calls when its customers need it.

First is the Model 125 site monitor. the power company uses 20 of them, says a company spokesperson. "Site monitors provide alarms and secure access to the switching equipment in our private network," he says. When you run your own network, there is no "other guy" (the carrier) to find and fix the problems. With its Model 125s, the power company knows when there is a problem and can fix it right away.

If the problem is with the switch (in the power company's case a Nortel PBX) calls can still be answered thanks to Gordon Kape's System 920 emergency bypass switch. If something happens to the switch (The spokesperson says most of the problems are with batteries and power supplies, ironically) a connection is automatically made between the local telephone company (Bell Atlantic) and specific phones in the call center.

This assures that customer calls are answered, no matter what. There are no special numbers customers have to call. The change-over is transparent to the caller. "It's a seamless backup," notes the spokesperson.

The power company's telecommunications strategies have proven so successful that the company has branched out to offer telecommunications services in addition to power.

The spokesperson used to work with telecommunications for the power company's call center, but now he is with a subsidiary, which offers long-haul communications services, including fiber optic services between two New England states.

The new subsidiary also uses Gordon Kapes site monitors to make sure everyone, including Big Three long distance carriers, has the information on conditions that they need.

ACD

The special services department of a major long-distance carrier wanted to buy a telephone system that would help handle special customer service requests in a way that reflects the company's own reputation as a leading high-technology vendor.

It used the PC-based automatic call distribution system from Applied Voice Technologies. The system is equipped with 40 trunks and Digital Display Sets for 16 agents.

Using the system, The Digital Display Sets have a four-line display which shows call status and other call information. This lets agents handle calls more efficiently and in general gives them more control.

For system managers, The system's Remote System Access lets them dial into the system from a remote location to monitor call activity, get reports and make programming changes.

ACD

One mid-Atlantic utility's eight call centers normally handle about 14,000 calls a day. During the worst of an ice storm that slammed its region one winter, its reps handled 47,000. Eight Nortel Meridian Call Centers (with ACD Max) helped them keep up with the deluge of calls.

The Meridians have a total of 460 ACD positions. They are linked with T-1s and use ISDN signaling. The utility uses NXX routing on their emergency 800 number to route calls to the right office. Networking calls between centers has reduced the abandon rate and the wait time.

ACD

A utility in the South wanted to accommodate a flood of customer calls after Hurricane Fran by switching from a failing old call center to a state-of-the-art new facility immediately following the hurricane, and a few days before the scheduled cut-over.

The new system was a Rockwell Spectrum ACD with UniverCTI for call routing and a CTI link to data.

Not only did the Spectrum ACD and the call center survive their trial by fire in the days immediately following the hurricane, but they were still handling higher-than-average call volumes four months after the hurricane.

Intelligent routing based on customer records stored in an IBM mainframe and automatic number identification (ANI) lets the utility bump customers on life support to the front of the queue during a power outage. This ANI information also lets the utility reduce talk time with each ANI hit. Their goal is to reduce talk time by 20 seconds per call on each ANI hit.

The center handled 500,000 in the first ten days after the storm.

ACD/VOICE MAIL

A Midwestern utility wanted to have callers spend less time on hold, to offer state-of-the-art customer service, to handle calls more efficiently. It did this with an ACD and voice mail from Executone Information Systems.

Using the system, the utility reported a 10.8% drop in the number of calls handled by live agents in the customer service department after the system was installed.

Callers to the customer service department now may choose how their call will be handled. A message board alerts callers to power outages or water emergencies. During high traffic calling periods, customers with non-emergency questions, such as questions about their bills, can leave a message and receive a prompt call back.

Less complex calls, such as meter readings, information about signing up for service and requests for investigator services can be left in a series of voice mailboxes.

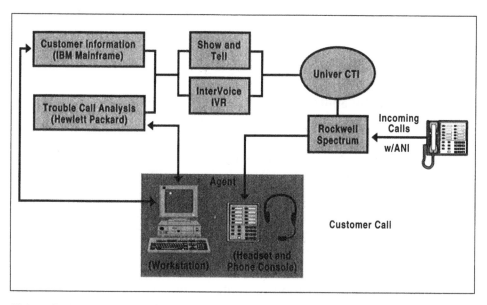

This utility's Rockwell's Spectrum ACD is linked to the company's IBM main-frame with Show & Tell and an InterVoice IVR system. The Spectrum talks to the UniverCTI for call routing and a CTI link to the data environment.

ISDN

This large, Midwestern utility company didn't go looking for an ISDN solution, but when they searched the marketplace for an automatic call distributor (ACD) to meet their needs, they found ISDN was the best solution.

The utility serves more than 400,000 homes in its service area. When their search for an ACD began, they had customer service representatives in all of their 19 offices, located throughout the region. They wanted to centralize their customer service operation in their headquarters.

They also wanted to use automatic number identification (ANI) to generate screen pops of customer information and be able into integrate the switch with the Windows software they already had. The answer was a Distributed Call Center from Teloquent.

"ISDN was not the driver behind the decision," says the utility's systems coordinator. "What intrigued us was the whole technical platform, which ISDN is a part of."

Today 35 customer service agents work out of headquarters call center. When things get busy, such as when a storm hits, personnel at the 19 other offices are able to handle the overflow of customer calls — thanks to the switch and ISDN links. This year the utility expects to boost the number of agents in the headquarters office to 60.

14

STAFFING

CATALOG

One popular children's clothing catalog company runs a family-friendly call center. That should be no surprise to fans of the company's bright, high-quality children's clothing.

The call center is open seven days a week from 5 AM to 9 PM, reports the call center leader. The center's 65 agents (or up to 132 during peak seasons) choose how many hours they would like to work. Schedules vary from 15 to 40 hours per week.

"We are really, really flexible with our staff," says the call center leader. She notes that the center attracts mothers who don't want to work full time, but want to sell a product they believe in and go home to their families and the rest of their lives.

This means the agents have some passion invested in getting just the schedule they hoped for. The catalog company's flexible policies are great for families, but they can be a little rough on call center management.

You won't find the call center leader complaining, though. A TeleCenter System call management software system from TCS makes juggling those flexible schedules a little easier, and it helps to calm some of those passions.

How can a piece of software calm passions? "It's pure science," says the call center leader. "It's not at all emotion-based. We can show people what this scientific thing says, and they can see that we don't need them at that time."

As most catalog companies know, even with the best scheduling software, pinpointing staffing needs can be difficult when you take the variables thrown into the mix by the US Postal Service. If a catalog drops a day earlier than you expected, it can throw your whole staffing schedule off.

That's why the call center leader relies on TCS' intraday performance reports (which are called daily activity reports at the catalog company). The reports are generated every two hours. Using this information the supervisor may decide to send people home or to call people in.

"The curves change depending on where the catalog drops," says the call center leader. For example, if the catalog is being delivered to homes on the East coast, then the bulk of calls will come earlier in the day. And if the catalog is delivered earlier than expected then the call volume predicted for the next several days will shift too.

The call center leader prepares for this by running "what if" scenarios with the TCS software.

It's not surprising at such a people-oriented company that the call center leader is less impressed by the sophisticated technology at her command than she is with the great support she gets from the TCS staff.

"Their help desk staff is incredible," she says. "They keep calling back until they know that the problem is solved. It's above and beyond what a vendor usually provides."

Catalog companies such as the catalog company feel like family to the customers who do business with them. It looks like this cataloger has found its match in this technology vendor.

HIGH TECH

OK, you have to decide how many agents must work on Christmas Eve. If you put too many agents on, they will be drumming their fingers and thinking evil thoughts as visions of sugar plums dance in their heads. If you put too few agents on, your customers will wonder why of all days, this is the day you can't help them get home quicker.

Would you rather rely on an educated guess or historical data?

A computer operates a support center for customers who buy their mainframe and mini-computers for critical, 24-hour applications. Of course the support center is open 24 hours a day, seven days a week, even on Christmas Eve.

The computer company also has a Remote Service Network. A modem attached to the company's computer dials the support center when there is a depositing hardware failure, a board failure, and in some cases, a disk failure. When this happens, a message appears on an agent's computer screen.

The computer company support center receives 275 to 300 calls a day. Those calls are handled by a staff of 14 agents, with four to seven agents working at any one time.

As for the Christmas Eve question, the support center supervisor, votes for the data. She's glad the computer company support center recently started using Telecorp Products' ACD Performance software.

"We used intuition before," she says. "What we wanted was historical data. It's hard to predict how many agents you'll need, or remember how it was the year before on Christmas Eve. We have to have someone here, but we don't want to have more than we need."

The call center management software package from Telecorp Products has answered that need. The supervisor says the reports give her the information she needs to accurately schedule agents.

The new software has also helped the supervisor discover the call center's daily rhythms. She now knows that call volume starts to rise about 10 AM and doesn't start dropping off until 4 PM.

The center also uses Telecorp Products' Agent Window and the supervisor says that the real-time information on the agents, color-coded on the computer screen so management can see at a glance what's going on with the agents is very helpful.

PUBLISHING

Getting 810,000 newspapers into the hands of readers every day is no easy task. And the task doesn't get easier when there are 1.2 million newspapers to deliver on Sunday.

With the help of about 350 independent home delivery distributors and 100 street sales distributors (who in turn employ about 5,000 carriers), this is exactly what one major newspaper on the East Coast does.

This complex operation generates 5,500 calls to the newspaper each day about delivery issues. Of these calls about 1,000 are from distributors.

The newspaper's Circulation Center employs about 90 customer service representatives to handle these calls. They do a good job. The service level in the center is 95% of calls answered within 20 seconds.

The reps do such a good job, in fact, that the circulation center staff serves as a kind of recruiting bullpen for the rest of the company. The center has a lot of turnover due to internal promotions.

Luckily, the center uses the TeleCenter System call center management software from TCS Management Group to handle their staffing and scheduling needs.

The TCS system has helped the center deal with that high level of turnover. It generates forecasts for up to 60 months at a time, taking into account historical call volumes, seasonal patterns, trends, holiday variations, day-of-week patterns, intra-day distributions and special events such as sales promotions or product introductions.

The typical turnover rate in the center is four to five people per year. This year, however, the center is dealing with a 20% turnover rate because of a new customer service division within the company. The TeleCenter System is giving the center the information they need to handle the situation.

For example, the center normally hires extra staff in the spring to cover for staff vacations. This year they brought in two groups of 10 reps based on the TCS software's call volume forecast.

"We've been pleased with the accuracy of the predictions," says the Circulation Services Manager.

The software also lets the circulation center know when they can best release people to other areas of the company without having a big impact on the center's service level.

With TCS the newspaper's staffing is so efficient, the center has been able to reduce the number of full-time equivalent positions by about 200 hours per week.

For the newspaper, getting those 810,000 newspapers into readers' hands may be a challenge. But for their readers, getting satisfaction from the newspaper's circulation center is always easy.

TRAVEL

A call center in Memphis handles calls for the gaming company's riverboat and Native American reservation casino operations. The riverboat casinos have call center needs that are a little different from the average casino because tickets are sold.

Ticketing creates sharp spikes in the center's staffing needs. People tend to visit these casinos when they have time off from work, says the company's Manager of Planning and Analysis for Teleservices Centers.

This means weekends are very busy, as are all holidays. The manager says the center's busiest time is the week between Christmas and New Year's, although holidays such as the Fourth of July are busy too.

The center handles about 2 million calls a year.

To make things a little more complex, the call center has several types of agents, which need to be represented in any schedule. There can be as many as 148 agents working at one time. The average is 70.

The casinos' call center has been using the TCS system since they split from their parent company last year. (They had used the TCS software as a department of that company's call center too.)

The TCS software is an integral part of the operation of this call center because of their unusual scheduling needs. "We couldn't schedule to this type of call pattern manually," says the manager.

15

HIGH TECH PRODUCT SUPPORT

CALL TRACKING SOFTWARE

Imagine your paycheck isn't coming. That's what is at stake when the customer of a payroll systems software company pick up the phone to call its technical support line. The company makes payroll and human resources software. When customers call around the first or fifteenth of the month, the problem could mean employees not getting their paychecks.

With the CustomerFirst call tracking system from Repository Technologies Inc., the company not only puts problems in queue, but it prioritizes "system down" calls so they can be handled first and everyone can get their paycheck. Using CustomerFirst, the company is able to respond to calls in two hours.

The company's service wins raves from its customers, but it is also part of the company's prestigious ISO 9000 certification, showing the company has met international standards of quality.

Best of all, CustomerFirst doesn't just help the company solve problems after they happen. It helps the find and resolve problems before they result in technical support calls. For example, after learning the top four reasons for support calls for its Payroll Year End programs, the company was able to resolve the problems and reduce support time.

CALL TRACKING SOFTWARE

The makers of network and communications hardware and software wanted to organize and formalize a "Corrective Action" process. This process was spread over many departments and included everything from safety violations to document audits. Requests were tracked by various means — through spreadsheets, PC databases or even in notebooks. Requests were shared with management monthly.

The company knew they wanted the software system they chose to be easy to customize, run on a variety of client platforms and connect with several other Windows-based applications.

It chose Vantive Support, Vantive Quality and Vantive HelpDesk from The Vantive Corporation.

Vantive Support processes and tracks customer issues and requests. Vantive Tools helps customize the system with interface screens and specialized workflow rules that let non-Vantive users get into the action through Groupwise E-Mail. Vantive Quality is used by the engineering quality group to get to the root of problems to help stop them before they generate a customer call. Vantive HelpDesk is used to support the internal help desk.

With the implementation of the system, the company was able to close more than 300 previously unresolved cases. They can whip through 2,700 up-to-date records in two to three seconds, where before it took 30 to 40 seconds.

More than 30 people started using the system when the company first implemented it. By the end of this year they expect that number to be over 250 people nationwide.

CUSTOMER SERVICE SOFTWARE

What if every one of your clients had the fate of the free world in his or her hands? That's the situation at the company that makes a correspondence management system used by members of the US House and Senate.

The company not only supports its own software for these congresspeople, but it also supports all the computer hardware and networks in the offices of several Representatives. In all, the company has over 200 support customers, each with the power to move and shake on Capitol Hill.

To handle this glut of VIPs, the company uses Support Magic from

Magic Solutions. Until recently, the company used a call tracking system it created in house, says the company's Director of Technical Services.

With Support Magic, the company can treat its customers like the VIPs they are. In fact, the company can offer even better service now that it can generate reports for each of its clients giving them a breakdown of the number and types of support calls they make.

The director raves about the flexibility of the SQL version of Support Magic. He says the company uses it to track items checked out from the company library and to track the training classes each support person has taken. He plans to expand the company's use of the system to include creating purchase orders and other automation projects.

DIRECT INWARD DIALING

A help desk outsourcer handles help desk calls for a Fortune 50 corporation using its Nitusuko America 384i phone system to route calls. Instead of using an automated attendant or a uniform call distributor (UCD) function, help desk callers dial individual agents' telephone number. The system takes calls on T-1 trunks and using Dialed Number Identification Service (DNIS) and a virtual extension button, the calls are routed to the proper agent.

The system uses Direct Inward Dial (DID) extensions to route incoming fax calls to an analog station port connected to a fax machine.

IVR

A major computer hardware maker is no stranger to automated technical support systems, but the AnswerDesk interactive voice response (IVR) system it developed with TTM & Associates has given them something it has long craved: feedback.

The hardware maker has 90 people on its technical support staff that are on the phone every day. They handle about 1,800 calls per day. The AnswerDesk server handles 2,880 calls during the same time.

For the directors of the electronic support team at the hardware maker, these great numbers mean nothing if they are not backed up with numbers they can use in management decisions. They demand accountability, statistical quantification of performance and quantification of customer satisfaction. They are getting all of this with the AnswerDesk system.

IVR

Industry has moved out of town. Smoke stops rising out of factory chimneys and the buildings decay. The home team is on a hot streak. The stadium is packed with adoring fans. The software is a huge success. Technical support is flooded with calls.

Urban decay, packed stadiums and even floods are all part of the incredibly realistic software titles from a California computer game maker. The software can predict and simulate all sorts of outcomes for a city designed by the user, but no one could have predicted the software's incredible success and the strain it would put on the company's technical support staff.

"When the DOS version of the game software was released we were swamped with calls," says the company's customer service manager. So many new buyers were installing the software that routine technical support questions ("Will this run on my '286?") came in a deluge.

The quick fix was to bring in temporary employees to aid the existing support staff. But when the increased holiday volume continued into May, it was clear that something more permanent had to be done.

Most of the software company's titles are entertainment software. People buy and use entertainment software in their free time — on evenings and weekends. "Eighty percent of our support issues are first time installation and running the program," says the manager.

Since so many calls were expected on evenings and weekends, adding more staff during business hours was not the only answer.

After extensive research into both their own needs and what was available in the marketplace, the software company found one solution in a 12-port IntelliSystem from Intellisystems. IntelliSystem is a combination of rule-based expert system software and an interactive voice response (IVR) system running on a PC platform.

IntelliSystem lets callers to the technical support center get solutions to their software problems no matter what the day or time.

Callers are asked to select their computer operating system and software product from a menu of choices. They are then asked to select a problem from a menu of problems found to be common for that operating system and software product. For example, a menu might list problems setting up, problems playing, and printer assistance.

If the problem is not one of the common problems contained in the system, during support center hours, the call is transferred to a technical support person.

When the caller selects one of the problem groups, the problem is broken down further. Callers selecting the "problems playing" [the game] prompt are then asked, "If your airports won't build, press one. If your seaports won't build or only build warehouses, press two...."

IntelliSystem remembers each answer and uses it to ask the next question, just as a technical support person working through on-screen expert system information would.

The session can be halted so a solution can be tried and resumed with another phone call. If no solution is found, the call can be transferred to a representative, who gets a screen of information along with the call outlining the caller's session with the automated system.

The software company was well prepared to create the knowledge base that drives the automated system. They had been using a call tracking system that gave them information on what products, platforms and problems callers most frequently asked about.

When the system was first installed, the manager had two of his senior technicians take home a list of the center's most common problems. Overnight they wrote all the solutions to those problems. Using those solutions, Intellisystems put together the automated system. From that minimal effort, the automated system was able to handle 15% of the calls coming in to the center.

"We've been growing the system from there," the manager says.

In fact, one of the things that pleases the manager most about the system is the ease with which new cases are added. "Training takes two days. They cover everything you need — in-depth." Once trained, it is a simple matter to add new solutions to the system. the software company does this constantly.

"When you listen to the system you will hear several different voices," he says. "That's because solutions are added on an ongoing basis. It's a team effort and we try to include everyone."

Maintaining the IntelliSystem has fit in nicely to the structure of the department. One person in the department holds the title of "automated systems administrator." This person is in charge of the in-house knowledge base and the automated system. the manager reports the person spends 15 to 20 hours per week on the IntelliSystem, doing what-if scenarios, and other planning and maintenance.

Adding solutions has increased the percentage of calls the automated system handles.

A significant bump in usage came when the software company changed the initial greeting on the technical support telephone line. At first automated support was the third choice on the menu, while speaking to a live representative was the first. Now automated support is the first option, a fax response system second, and the live representative the last option given.

The offload, or amount of calls handled by the IntelliSystem, has increased from that initial 15%, to 30% last October, 35% in November, 42% in December, and averages over 40% so far in 1996.

Using the automated tech support system has let the department keep up with a quickly expanding product line.

"We have added new products like you've never seen," says the manager. "It's a really aggressive release schedule. And it has been a real benefit having the system in place when these releases occurred. We've been able to address known issues right from a software's release."

One of the ways the technical support department has been able to do this is with a staffer acting as a product liaison with the product development people. This person collects technical information about a new software product and puts it into the system before the product is released.

The department is able to provide 24-hour, seven-day-a-week automated help on a product from the first phone call it receives.

The department is constantly fine-tuning the system through an analysis of reports and feedback from representatives.

The manager has asked the reps to do a little detective work when a call is sent to them from the automated system. Was the solution in the system? Where did the caller go astray?

If a problem is found prompts are reworded or the pace of delivery is slowed until the path to the solution is clear. If the solution was not in the system, that fact is noted.

A new problem and solution is added when call tracking statistics demonstrate it has become a significant percentage of the overall calls to the center.

The information generated by the IntelliSystem and other technical support statistics are used to solve product related problems and improve documentation.

"The majority of the calls we get are for problems external to our software," says the manager.

When the department's reports reveals a problem, they may call the hardware vendor or put more information into the software to eliminate the problem before it affects the customer.

The software company had a three step plan for their automation product. The first step was to offer customer service after hours. The second step was to offer automated help during the day.

One of the most obvious results of using the IntelliSystem is that, for most of the year, the department has been able to cut down its hours of operation. They once staffed the phones from 6 AM to 6 PM to handle all of the calls they received. They now operate from 8 AM to 6 PM, except for the busy season surrounding Christmas.

The manager says the reduction in hours was due largely to the success of the IntelliSystem.

While the software company has reaped many tangible benefits from using the IntelliSystem, such as reducing call costs and trimming their hours of operation, the biggest benefits of the system may well be intangible.

The customer who buys one of their software products on a Saturday morning and is able to find out on Saturday afternoon why her airports won't build is likely to have very warm feelings toward the company that has made having so much fun so easy.

"The last thing people want to do is pay for a call when they have just bought software," says the manager. "We are conscious of that."

Most of the entertainment software industry, he notes, offers technical support through toll (caller pays) numbers only. The few vendors who offer technical support on a toll-free line offer a lot of add-on products and use their technical support line as a selling medium.

In spite of the fact that they don't have a lot of add-on products to sell, the software company is experimenting with automated support through a toll-free number to offer superior customer service.

The deciding factor is how much money the company saves on each automated technical support call. Once the manager could show the significant savings the company is getting with the automated system, getting the go-ahead to use a toll-free number was easier.

PROBLEM RESOLUTION SOFTWARE

This big-name PC maker has a vision: "to effectively and efficiently handle every customer event from conception to completion."

It's an ambitious vision, and one most call centers, in high tech product support or not, can relate to. To make this vision a reality, the PC maker has gone beyond call center management, beyond telecommunications, beyond computing — into the realm of philosophy.

That a help desk software package from Inference could dovetail into the philosophy behind this high-minded effort certainly speaks well of the system itself.

After speaking with the PC maker's Manager of Customer Support Operations, it's obvious that a lot of thought has gone into the big picture — not just the nuts and bolts of getting from here to there.

"We had to analyze our resources to accomplish that vision," says the manager. "Inference was just one tool in this. It is really a process reengineering effort. Our philosophy is you have to have the processes in place that feed the technology and tools. The infrastructure drives the technology and tools, but the most important thing is to ask, 'How would this change effect our customer?'"

The PC maker has three call center locations in two states. The centers have two functions, traditional customer service and technical support. The technical support function is found at all three sites. Customer service is at two.

The PC maker won't reveal the number of agents or the number of calls received in the centers. But the manager says the operation is "pretty large."

Until recently, the centers were using several different systems to manage information about customer events, some developed in house, others purchased as turn-key systems. As the corporation grew exponentially, says the manager, this hodge-podge of systems no longer met their needs.

An in-depth study, begun about a year ago, revealed what needed to be done to deliver information to the right place, and to manage that information. "The study looked at the issues from both a process standpoint, and a tools and technology standpoint."

The PC maker decided what they needed was a way to share solutions to those 80% of customer problems that were most common and already solved. Key-words and a series of questions-and-answers would lead the agent to the proper solution.

If no solution is found, agents have the option of entering an "unresolved case" into the system. When technicians solve the problem, they feed it back into the system. In this way, the PC maker's store of knowledge is always growing.

One key factor in developing and rolling out the Inference software, was the involvement of the agents (called Customer Support Representatives and Customer Support Technicians) from the very beginning of the project, says the project manager for the implementation.

What is the ultimate goal of both the PC maker's philosophy and the use of the Inference system? Says the customer service manager, "We want to achieve a competitive advantage through excellent customer service." A sentiment any call center can relate to.

16

VIRTUAL CALL CENTERS

CABLE

A big name cable company in one of the nation's largest cities has big plans for an integrated, intercity call center. And they need big plans. This cable company serves 730,000 subscribers with 400 agents. They handle a minimum of 20,000 calls per day.

Their goal is to have two integrated call centers, each served by a Rockwell Spectrum ACD, says the vice president of customer service and marketing for the cable company.

There are two call centers, each located in the part of the city it serves. The idea is to overflow calls from one center to the other when the traffic at one is high.

But the plans don't stop there. When complete, the system will route calls to specialized agents based on callers' response to a voice response unit (VRU) message.

Automatic number identification (ANI) will trigger an automatic screen-pop of customer information on the terminal of the best-suited rep. Outbound calls will be made from an integrated predictive dialer. When things on the inbound side get busy, the outbound agents receive inbound calls with the push of a button.

The call center will be largely paperless.

Many pieces of this plan are already in place. A Rockwell ACD was installed in one call center. The other ACD should be installed in a few months.

Already callers to the first center are greeted by a VRU and routed to an agent who is specially trained to answer his or her question. All agents are cross-trained to handle all call types, so the ACD can select the most-qualified agent for each call, but callers don't have to wait for a single agent.

The VRU also collects the telephone numbers callers enter using their touchtone keypads. The number is used to deliver a screen of customer information to the agent.

A Windows-based software system is helping the cable company achieve a paperless office. The VP says the new software makes interacting with the company's billing system easier. And having on-line information makes customer service more consistent.

One of the reasons the Rockwell Spectrum ACD was selected, she says, was Rockwell's experience with pulling a sophisticated system like this one together.

The VP has no doubt that the system is already helping to improve customer service, but in her eyes technology is just the vehicle. Having "well-trained, informative people to answer questions" is the driving force.

The cable company's new call center technology will help their agents, their driving force, work more efficiently, with better information.

CATALOG

At one high tech/luxury catalog, total call coverage is as easy as one, two, three. The company's main call center is located on the West Coast. And an outsourcing company located in the East handles those off-hours calls in the middle of the night.

In between is a small call center located at the company's distribution site. It smoothes the transition between the main call center and the outsourcer.

The Call Center Supervisor at the distribution center explains that the call center handles about 1,000 calls a day with a Cintech Cinphony ACD.

During the holiday peak calling times, the distribution center center has as many as 17 agents. During the rest of the year, nine agents is about average, says the supervisor. (Father's Day is the second busiest time of the year, but the West Coast center staffs up to cover that one.)

A network routing feature from AT&T allocates percentages of the total volume of incoming calls to two of the three call centers depending on the time of day. (Only two centers are available at any one time.)

When the distribution center starts cranking at 8 AM, it gets the lion's share of the incoming calls, with the East Coast outsourcer serving as backup for late starts.

At sunrise on the West Coast, that call center starts taking calls, so as the day goes on, more and more calls are allocated to the West Coast center.

In late afternoons and on weekends, the West Coast and East Coast call centers provide call coverage.

One of the functions of the distribution center call center is to provide back-up call handling for the West Coast site, but the supervisor suspects that the West Coast center has covered for the smaller call center as frequently as the smaller call center has covered for the main location.

The supervisor uses an agent's average talk time to manage agents to higher productivity and sales. One disappointment she has with the system is that while it provides a average talk time per group report, the statistics on individual agents have to be calculated by hand or through a computerized spread sheet.

In all, though, the supervisor is pleased with the Cinphony ACD, the dedication of the agents at the distribution center call center and her role as supervisor.

"Our customers are typically very pleasant to deal with on the telephone," the supervisor says. "It's a fun job to deal with them."

TECH SUPPORT

After ten years of revenue growth a leading provider of computer adapter cards needed to expand and enhance its technical support services. Routing calls through its PBX had become unmanageable.

Installing a CallCenter ACD from Aspect gave the company much more than a way to route calls efficiently, however. Sure, the CallCenter ACD kept up when the company expanded from 24 to about 130 technicians, and when the number of calls handled topped 2,000 per day. But it did so much more.

Today, as many as 10% of the company's technicians work from home. Multiple call centers, two in California and one in Colorado, form a virtual call center that has opened new doors for the company.

With the virtual call center, the company doesn't have to rely on the stressed Silicon Valley labor market for all its technicians. The Colorado facility offers another pool of highly skilled technicians. And many techs in from its Silicon Valley headquarters are thrilled with the opportunity to move to Colorado.

The company can give its techs the option of working at home or at a remote office because the CallCenter ACD gives them transparent networking across the call centers, the ability to make changes as needed and reporting that includes all the stats needed for important decisions.

TECH SUPPORT

Using remote agents is something new for most help desks, but the technical support folks at this software company have been working from home for some time. Four of the help desk's 10 agents work from home.

Back when the company was using a simple, mainframe based system to track its calls, they linked these agents through a terminal session over the Internet. The company has gained many benefits by using the client/server based Applix help desk software and now that it uses WebLink, those benefits are available to its remote users too.

After using the Applix software for just a few months, the company (which makes database software for college administration) could use the information entered into the system for trend analysis. The company now knows who is calling when and why. For example, there is a flood of financial questions in July, the end of the fiscal year for colleges. "We knew we were hit hard, says the Senior Systems Manager, "but we didn't know how hard until we had this information."

Support staff have been entering information about problems and solutions since the company started using the system about a year ago. Getting the techs to enter the solution into the system required a new way of thinking, says the manager, but now the company has started to reap the rewards of their work.

With all the information entered into the Applix system, techs have previous solutions to help them solve problems, management has information on customers' most common questions for better training and everyone has a better idea of what to expect based on past experience.

The next step is to open up part of the system to customers over the Internet using the WebLink product.

UTILITY

An electric utility that handles 10,000 calls daily wanted to, "provide customers with the absolute best possible service..." through a virtual call center and intelligent call routing.

It used Lucent Technologies' Definity G3r phone switch (with Expansion Port Network) and a Conversant IVR system

Using the system, the utility's virtual call center links 110 agents in several locations, including 80 agents in the primary call center, two remote nodes, six home agents and another site in the southern part of the state.

Expansion Port Network (EPN) a feature of the switch, along with T-1 circuits and a SONET fiber optic network let much of this happen. The utility is very pleased with the responsiveness and productivity of its home agents, who are connected to the main switch through digital extender units, and plan to expand the number of home agents.

Intelligent routing lets customers prioritize their own calls through the Conversant IVR system. But if someone selects "power outage" when the company's database shows service was cut off because they didn't pay their bill, the call will be directed to a collections rep instead.

With outages, the IVR system can collect outage information and deliver an estimated time the power will be restored. Messages played in queue are also tied to caller information — and the time of year and if there is an outage.

UTILITY

A small utility that handles over 12,000 calls a month, wanted to improve customer service, both to walk-in and telephone customers, without adding additional reps. It used a Siemens Rolm 9006 Model 80 PBX/ACD.

Using the system, the utility consistently maintains a service level of 97% of calls answered within 20 seconds, with about 11 agents, five of whom work primarily with walk-in customers.

The Manager of Customer Service reports that in one month earlier this year, the average speed of answer was seven seconds over-all. It was even better during the hours the call center was open. (A dispatch department handles calls when the customer service call center is closed.) Since the call center was created last year, its average speed of answer has been eight seconds.

Before the call center was created, customer service reps had to juggle walk-in customers and telephone calls. Now six dedicated telephone reps handle calls, while other reps primarily deal with walk-ins.

Customer service reps in two remote locations are able to log into the system and handle calls when they are not serving walk- in customers.

The manager feels this feature makes a big difference in the level of customer service the utility is able to offer its customers.

The manager says, "We have a real good group of customer service reps who have been with us a long time and do a good job." Perhaps it is this confidence in his reps that leads him to focus on how the team is doing as a whole, and put less weight on the statistics for individual reps.

TABLES

FUNCTIONS BY INDUSTRY

	Banking	Catalogs	Financial Services	High Tech	Package Goods	Retail	Utilities	Cable
Billing and Credit	53							5
Collections			18				19	16, 17
Fundraising								
Help Desk	37			49-51		53	39, 40	
Order Handling		24, 59, 62-67, 109, 124		110	25, 67	64, 70		59
Sales	23, 77			82-87		89		77-80
Customer Service	53, 95-98		98, 99	13	101	102	27, 103-106, 127	98
Product Support		115-122, 125-126						
Information Services	41-44		43, 45	46		47		
Other	11	55					12	

Collections Agencies	Government	Fund raising	Publishing/ Entertainment	Health care	Service Bureaus	Education	Travel
6, 7				7			
11, 12, 15, 16	18						
		31		30	29, 31, 32	34, 35	
	38						38
			67, 68, 111		60, 69		70-75, 112
			87, 88	80-82			89-93
123	99			100			102
			44, 45	13, 46			47
				25, 56			

TECHNOLGY BY INDUSTRY

	ACD	IVR	Predictive Dialing	Call Center Management Software	CTI	Help Desk Software (Knowledge based)
Banking	43, 44	41-43, 97-98		5, 95-96	23	37
Catalogs	59-63			63-66, 109	23	
Financial Services	98	45	6, 7, 18			
High Tech	13, 100	13, 117		50, 86, 110		46, 49-50, 121
Packaged Goods	67, 101				25	
Retail	47, 70	102				53
Utilities	19, 104-106	105		39	27	39
Cable		59	16, 77-80	16, 98		
Government	99		18			
Fundrasing/ Nonprofit			30-34			
Publishing/ Entertainment	67, 87-88	67, 68		111		
Health Care	7, 46, 100	81	81	100	24	
Service bureaus	11, 69	11, 13	15, 16			
Travel	70-72, 89	47, 72,	91, 102	73-75		

Problem Management/ Customer Service Software	Fax	Speech Rec	Monitoring Systems	Sales Software	Virtual Call Centers	Other
77	44	44				11, 12, 53, 96
			55		124	66
				98		
115-117				82	125-126	117
				89		
					126-128	103, 106
					123	
38						
						35
88		44, 45				
			56	80		
						11, 12, 69
38						75

INDEX